The One Song

Unified Mechanics of Emanation

A Philosopher's Support of Strings

By

B. K. Bodish

~ The Presentation of The Ideal Field Model ~

A MASS MEDIA PUBLICATION

The One Song
Unified Mechanics of Emanation
A Philosopher's Support of Strings

Published by:

MASS MEDIA
PUBLICATIONS DIVISION
PO BOX 1411
Cranberry PA 16066

ISBN 978-0-9988148-2-7

The One Song

Unified Mechanics of Emanation

A Philosopher's Support of Strings

By

B. K. Bodish

~ The Presentation of The Ideal Field Model ~

0. The Ideal Field Model

There is a shape uniquely described by Marko Rodin. Presented in this primer are some *basic* concepts of his and others' work, along with the author's, all compiled into one unified model with far reaching implications. The goal of this work is to inspire readers to examine in far greater detail the concepts to be briefly presented herein and apply them to their field, from nanoengineering to plasma cosmology.

This shape is the torus:

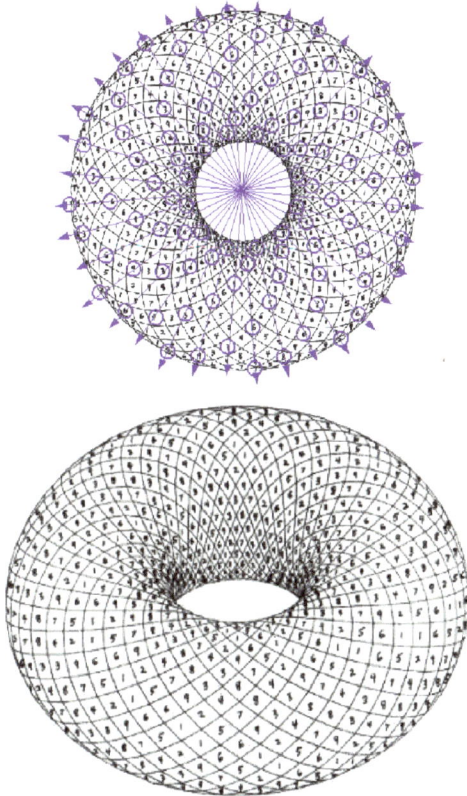

Keep in mind this is a 3-dimensional (at least) image.

The above images are Marko Rodin's Torus number pattern. Mr. Rodin correlates the torus to a natural pattern inherent in numbers. Comparing the numbers 1-9 to themselves reveals a pattern which extends into countless systems.

7

This pattern can be visualized with the following diagrams by Marko:

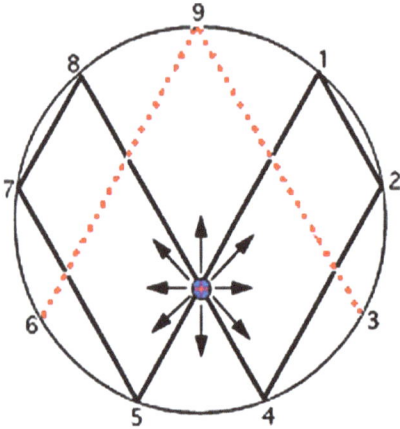

The numbers 1, 2, 4, 8, 7, 5 on the black line (of the first diagram, left) are connected by doubling, starting from one. Any product that is over 9, the numbers of that product are added together. For example: 8 doubled is 16, but 1+6 equals 7. Notice that halving the numbers does likewise, but in reverse. Simply add together the numbers which make up the halved decimal, and the same pattern is revealed. (The middle chart is doubling without reduction, while the bottom chart is halving without reduction)

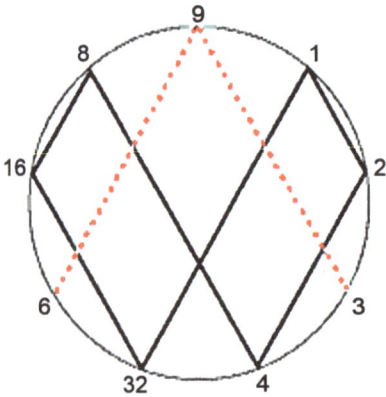

Additional self interactions of the numbers 1 through 9 appear on the tables on the next Page. First, on the left side of the first table, the rows and the columns are multiplied together, and again, any product over 9 will be added together to achieve a single digit. For example 7X7=49 but (4+9) = 13 and (1+3) = 4 so 7X7 = 4 in this system. (Notice the colored columns and how their sequences are mirror images)

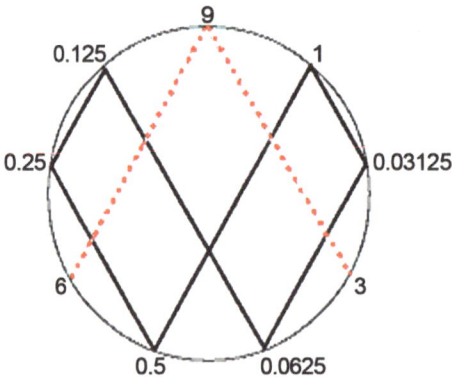

Next, the right side is the division table, and filling it in is a bit tricky because of (seemingly) complex decimals. First, solve for what you can, and then check the results by using the mirror pattern that was previously discovered in the multiplication table. You will notice another pattern, which is the number of spaces in between 1,2,3,4 etc. Notice that row one has no spaces, row two has 1 space, row three has 2 spaces, and row four has 3 spaces, etc. This assists in proving the pattern. A great deal can be learned by constructing this division table from scratch without looking at this finished table. Only in this way will it be clearly understood and reveal meaningful insights.

Multiplication and division:

	1	2	3	4	5	6	7	8	9	1	2	3	4	5	6	7	8	9
1	1	2	3	4	5	6	7	8	9	1	2	3	4	5	6	7	8	9
2	2	4	6	8	1	3	5	7	9	5	1	6	2	7	3	8	4	9
3	3	6	9	3	6	9	3	6	9	3	6	1	3	6	2	3	6	3
4	4	8	3	7	2	6	1	5	9	7	5	3	1	8	6	4	2	9
5	5	1	6	2	7	3	8	4	9	2	4	6	8	1	3	5	7	9
6	6	3	9	6	3	9	6	3	9	6	3	2	6	3	1	6	3	6*
7	7	5	3	1	8	6	4	2	9	4	8	3	7	2	6	1	5	9
8	8	7	6	5	4	3	2	1	9	8	7	6	5	4	3	2	1	9
9	9	9	9	9	9	9	9	9	9	9	9	3	9	9	6*	9	9	1

Addition and Subtraction

	1	2	3	4	5	6	7	8	9	1	2	3	4	5	6	7	8	9
1	2	3	4	5	6	7	8	9	1	0	1	2	3	4	5	6	7	8
2	3	4	5	6	7	8	9	1	2	1	0	1	2	3	4	5	6	7
3	4	5	6	7	8	9	1	2	3	2	1	0	1	2	3	4	5	6
4	5	6	7	8	9	1	2	3	4	3	2	1	0	1	2	3	4	5
5	6	7	8	9	1	2	3	4	5	4	3	2	1	0	1	2	3	4
6	7	8	9	1	2	3	4	5	6	5	4	3	2	1	0	1	2	3
7	8	9	1	2	3	4	5	6	7	6	5	4	3	2	1	0	1	2
8	9	1	2	3	4	5	6	7	8	7	6	5	4	3	2	1	0	1
9	1	2	3	4	5	6	7	8	9	8	7	6	5	4	3	2	1	0

As one can see, the addition and subtraction tables, albeit simple, also continue a uniform pattern. However, the focus shall stay on the multiplication and division table. Notice again the mirroring pattern of the numbers, now across from each other on the diagram below, placed by the number that headed the solid color column or row in the above table:

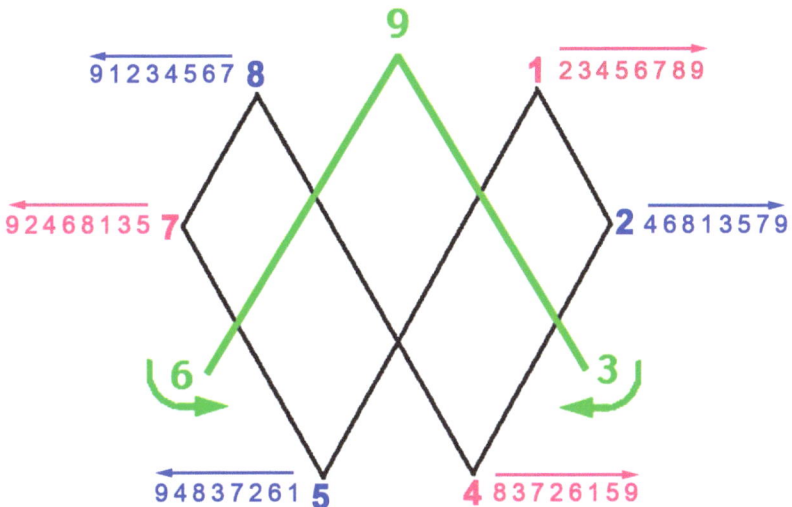

The doubling path is followed; 1, 2, 4, 8, 7, 5, and back to 1, making the "infinity" or "figure-eight" sign. Notice that even though the sequences themselves are mirrors, that the numbers in those sequences are *also* individual mirrors, regardless of the sequence that they are in. For example: 1 is *always* opposite 8. It then becomes clear that 1,4,7 and 8,5,2 and 3, 6, 9 are special groups of numbers (colored pink, blue, and green respectively on the previous diagram). Marko calls these "Family Number Groups". These groups are always separated by thirds. For example: family 1,4,7: $\underline{1}+3 = 4$, $\underline{4}+3 = 7$, $\underline{7}+3 = 10$ (1+0) which is $\underline{1}$, returning the cycle to the start position of the family. The 8,5,2 family has the same relation ship: $2+3 = 5$, $5+3 = 8$, $8+3 = 11$ (1+1) is 2, which is the starting point.

Notice it also works with 6, but in reverse. Thus the 3 and 6 are related, much like an oscillating cycle. The oscillation can further be viewed by doubling the 3 and 6: 3 doubled is 6, 6 doubled is 3 (12=1+2), 12 doubled is 24 which is 6 (2+4) etc. Constant oscillation from 3 to 6, the nine being an unseen boundary condition.

The 9 is eternally whole and self similar and emanates in multiples of 7. As any number divided by 7 equals the sequence 5,7,1,4,2,8 which is the above 1,2,4,8,7,5 pattern in reverse with the 1 and 8 positions swapped (hinting at 1 and 8 as being a pivot point). 9 is also the control of polarity, if it is positive then 3 and 6 will both be negative. For example: -3, +9, -6, +6, -9, +3, -3, etc. Lastly, all of the 9's on the model are vortices on a smaller scale than the model itself, showing that the entire model *is* a "9". This is one example of the model's self similar recursion.

It is thus possible to think of the 3,6,9, as an "Energy field" and the 1, 4, 7, and 8, 5, 2 as "matter", or the "aspects" of matter. Any two numbers from each "matter" groups will always equal the field. Example: 7+8=15 which is 6. (1+5) and 7+5=12 which is 3 and 7+2= 9, all field numbers. Any 2 numbers within one "matter" group equal a number in the opposite group. Furthermore, any two adjacent numbers surrounding a "matter" number, when multiplied together, equal that center number multiplied by itself. (see chart pg. 12) Example: 2 is surrounded by 1 and 4. 2X2 = 4 and 1X4 =4. Another: 4 is surrounded by 8 and 2 4X4=16 which is 7, 8X2 = 16 which is 7 etc.

With these endless patterns, soon "motion" of the numbers and multi directional symmetry become apparent on the diagram. It becomes clear that the numbers that make up the diagram are precisely placed in this symmetrical pattern by each number's unique "signature" which is its natural immutable pattern.

Next all of these "signatures" will be placed on a three dimensional diagram to understand them even more clearly. To do so, simply take the sequence from the diagram and lay it out on a grid. Follow the diagram in a "figure-eight" to get the 1,2,4,8,7,5 doubling pattern. It is this pattern which is drawn onto graph paper, one line going one way, the next going the

10

opposite direction separated by the 396693 pattern as shown below: (It is also interesting to note that the music scale doubles in frequency at each C note, therefore the exact same pattern emerges in the music scale: 1,2,4,8,7,5)

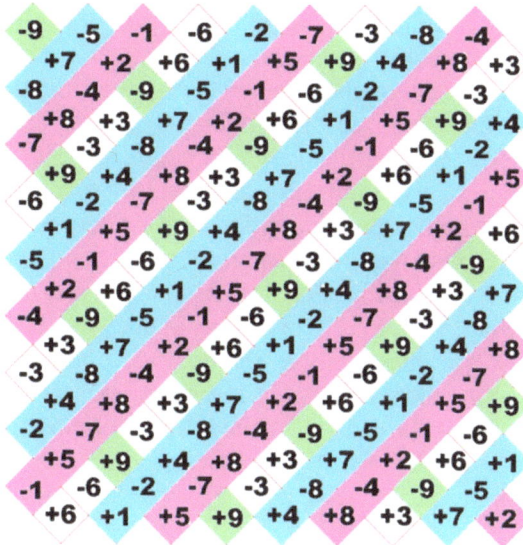

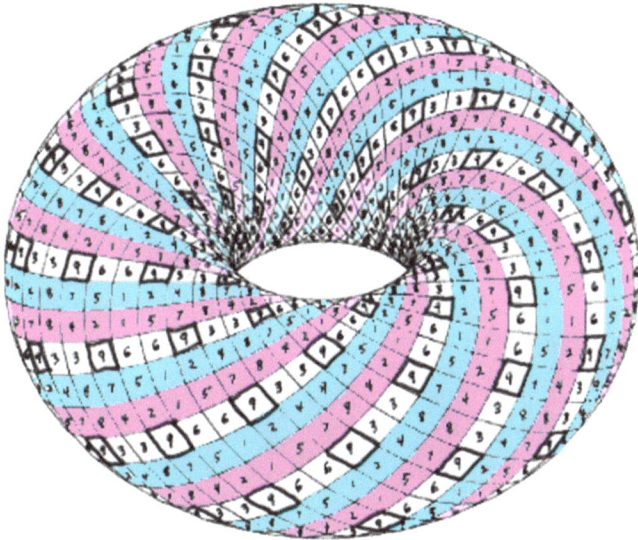

-9	-5	-1	-6	-2	-7	-3	-8	-4
+7	+2	+6	+1	+5	+9	+4	+8	+3
-8	-4	-9	-5	-1	-6	-2	-7	-3
+8	+3	+7	+2	+6	+1	+5	+9	+4
-7	-3	-8	-4	-9	-5	-1	-6	-2
+9	+4	+8	+3	+7	+2	+6	+1	+5
-6	-2	-7	-3	-8	-4	-9	-5	-1
+1	+5	+9	+4	+8	+3	+7	+2	+6
-5	-1	-6	-2	-7	-3	-8	-4	-9
+2	+6	+1	+5	+9	+4	+8	+3	+7
-4	-9	-5	-1	-6	-2	-7	-3	-8
+3	+7	+2	+6	+1	+5	+9	+4	+8
-3	-8	-4	-9	-5	-1	-6	-2	-7
+4	+8	+3	+7	+2	+6	+1	+5	+9
-2	-7	-3	-8	-4	-9	-5	-1	-6
+5	+9	+4	+8	+3	+7	+2	+6	+1
-1	-6	-2	-7	-3	-8	-4	-9	-5
+6	+1	+5	+9	+4	+8	+3	+7	+2

0.

(Image 0. Is the Field Model Not in its Ideal Form, presented as such for clarity)

The 1,2,4,8,7,5 pattern is colored blue and pink to show the equal yet opposite nature more clearly. The 9's have also been highlighted for visual clarity. Connect the left side of this chart with the right side by making a tube, next connect the top and bottom of the tube to make the Torus. Notice how the pattern continues without interruption.

Now one has a 3 dimensional model of these inherent "signatures" or "true identities" of the numbers 1-9, which are the ONLY numbers. In a sense, one now has a model of the 9 forces of the universe, and moving forces on a torus can be described as a vortex. This shall all be examined in much greater detail throughout this paper.

Continuing with the most notable symmetries; take any one tile on the torus, and disregarding charge, the numbers directly surrounding that tile will always total 9. With charge accounted for, the surrounding tiles plus the center always equal 9. Again disregarding charge, any size block describes specific characteristics of the numbers and their interplay with each other. For example: on the below diagram notice the two blue squares, one is 2X2 and the other is 5X5. Odd number squares always expand upon the nature of their center number. All even blocks, such as the 2X2 block, always equal 9. The 5X5 block equals 9 on every level surrounding the center, however, 9 plus the center always equals the center number. Now take charge into account and explore the more profound patterns on your own.

The positive and negative nature of the 9 can be examined in the 2 grey squares. Observe how they are mirror images of each other. Now observe the Red line separating the positive and negative polarities, multiplying the numbers across from each other on that line always yields 1. Example: 5X2=10 which is 1 (1+0), 7X4=28 which is 1 (2+8) etc. But when added across the red line, it is always 7 & 2 oscillating. Rodin calls this the Boundary condition.

Now notice the yellow squares, and how they are the family number groups aligned on every third band, and remember that each number in a family is separated by 3, ($\underline{1}$+3 = $\underline{4}$+3= $\underline{7}$). Also, from the bottom left yellow tile up the left side is the number sequence 1-9 to lay out the bands of the grid. Lastly, note that the total number of tiles is 162, (9X18):

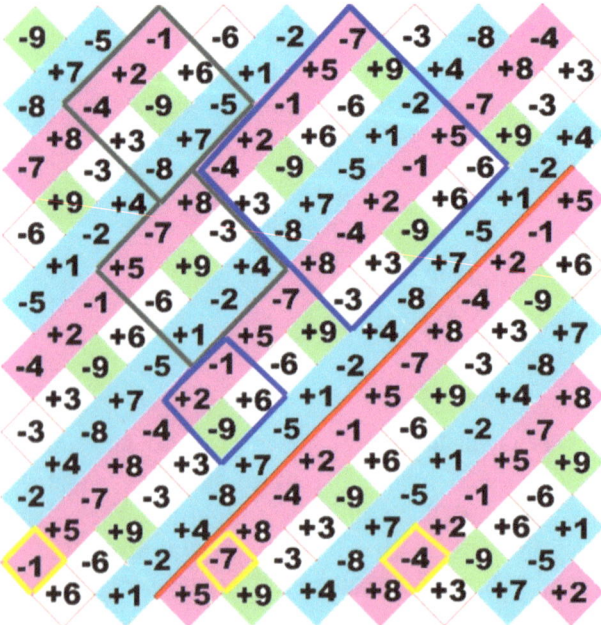

162 / 9 = 18.
162 / 6 = 27.
162/ 3 = 54.

All mirror image numbers on the doubling diagram.

1 + 6 + 2 = **9**

```
-9  -5  -1  -6  -2  -7  -3  -8  -4
+7  +2  +6  +1  +5  +9  +4  +8  +3
-8  -4  -9  -5  -1  -6  -2  -7  -3
+8  +3  +7  +2  +6  +1  +5  +9  +4
-7  -3  -8  -4  -9  -5  -1  -6  -2
+9  +4  +8  +3  +7  +2  +6  +1  +5
-6  -2  -7  -3  -8  -4  -9  -5  -1
+1  +5  +9  +4  +8  +3  +7  +2  +6
-5  -1  -6  -2  -7  -3  -8  -4  -9
+2  +6  +1  +5  +9  +4  +8  +3  +7
-4  -9  -5  -1  -6  -2  -7  -3  -8
+3  +7  +2  +6  +1  +5  +9  +4  +8
-3  -8  -4  -9  -5  -1  -6  -2  -7
+4  +8  +3  +7  +2  +6  +1  +5  +9
-2  -7  -3  -8  -4  -9  -5  -1  -6
+5  +9  +4  +8  +3  +7  +2  +6  +1
-1  -6  -2  -7  -3  -8  -4  -9  -5
+6  +1  +5  +9  +4  +8  +3  +7  +2
```

Note that Life itself (Cell growth) happens by division, which creates the 1,2,4,8,7,5 sequence. This sequence makes a vortex (as shown), which can be observed throughout nature: 1. All whirl pools, hurricanes, and tornadoes are vortices in different media, 2. Ram horns, nautilus shells, and plants grow in spiral vortices. 3. DNA is a spiral vortex 4. String theory suggests that the true structure of atoms may be toroidal energy vortices. 5. The magnetic field of planets and stars are toroids (shown in detail later).

a. noaa.gov

b. Spiralwishingwells.com

c. Richard Giddins, wikipedia.org

d. nasa.gov

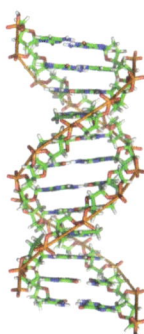

e. "JenaBug", wikipedia.org **f.** Richardwheeler.net **g.** Jim Chen, wikipedia.org

13

1. The Spiral Force

The point of describing the torus in such a way is so that it can be used as a model for force in nature. As quantum physics and plasma cosmology are now plainly showing, electromagnetic waves are at the root of all phenomena. Therefore this chapter will show how electromagnetic waves follow this Ideal Field Model so that it can be used to model any function of nature.

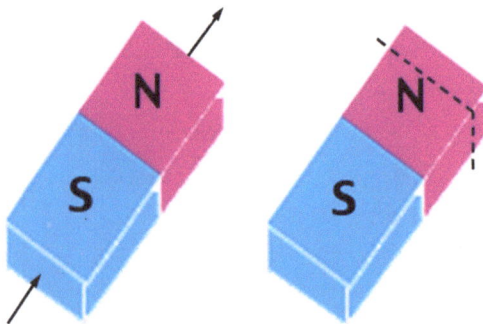

1.

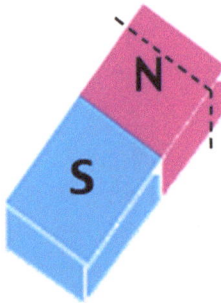

2.

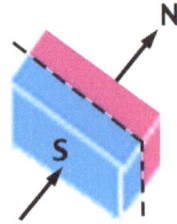

3.

Image number 1 is the basic accepted idea of a magnet. The basic force around such a magnet is shown in images 6 and 7. If a small slice of that magnet is taken (image 2.), the slice instantly polarizes and possesses the same qualities as the parent although on a smaller scale (image 3) this is called recursion. The slice has a meridian that is perpendicular to its axis, aligned as its parent. The axis is defined by a ray attaching the two extreme poles. If two such slices are put together, axis to axis, an imitation of the parent magnet is recreated (image 4).

However, when the magnets are combined *side   by   side*, (image 5) they attach, but the final field is much different from the circular field of image 6 and 7 (see image 14). It is in this arrangement where the relation to the torus becomes clearer.   To understand it better, the idea of north and south are better thought of as a vectors (directions of motion) as opposed to "charge".

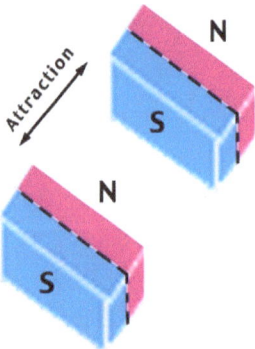

4.

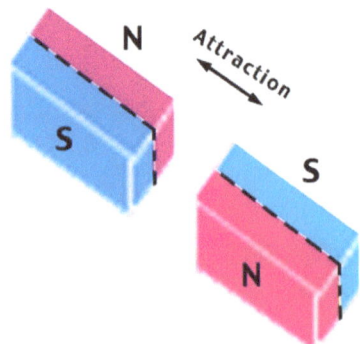

5.

Image 6 shows the vectors commonly thought of, and image 7 is showing the same vectors in classic iron filings.

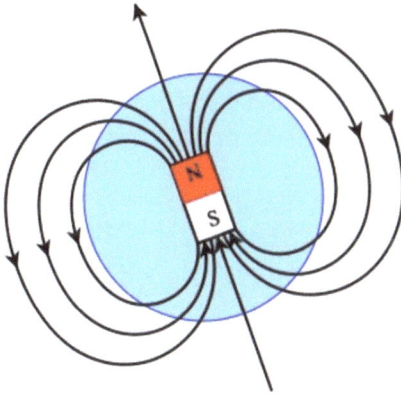

6.

TStein, wikipedia.org

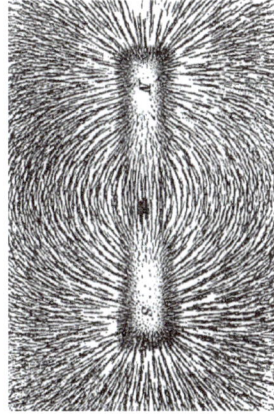

7.

Newton Henry Black (1913)

The vectors above depict a motion similar to two gears meshing together (image 8). Yet, all of these images are merely 2-dimensional. To draw a complete parallel to the 3-Dimemsional Ideal Field Model, vectors need to be plotted for the *sides* of the magnets as well (image 9).

8.

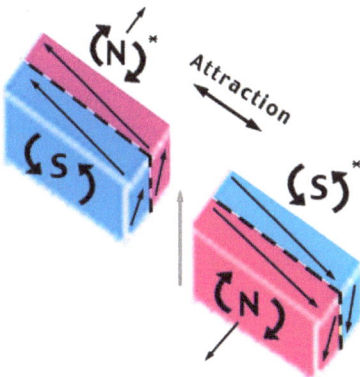

9.

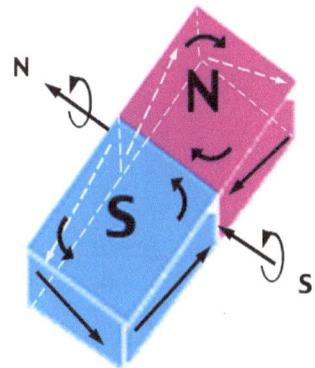

10.

9. (Asterisk means "relative to point of view")

15

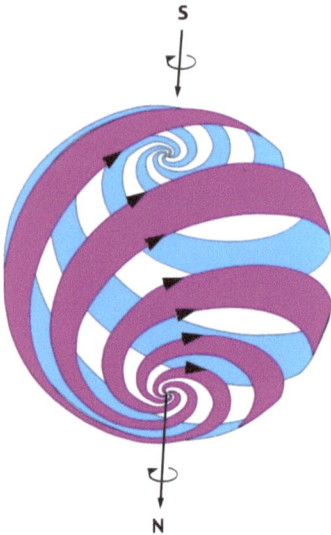

When vectors are plotted for the sides, the spiral vortex is clearly revealed and their interacting behavior becomes predicable. Image 11 shows a spiral sphere as a "stylized" representation of the final vector field that is generated after plotting the vectors in 3Dimensions. This is the "*Ideal*" form of the torus field model. Understand that a sphere with an axis *is* a torus, because the axis is a hole, however small.

Note: Image 11 is not an exact model (such as the specific angles of the vectors), it is being considered here for sake of simplicity in modeling.

11.

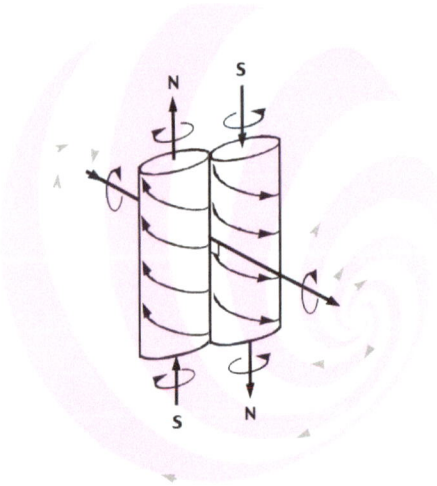

Image 11 and 12 detail the 3D vectors of the two axis scenarios: axis *to* axis, and axis *by* axis respectively. Now the interaction that occurs when vortices are *side by side* can be better understood, and one can see why the final fields of the two are so different (see image 14 again). While image 4 aligns to form one single axis, image 5 has two axes. A "phantom" axis then attempts to form perpendicular to these two existing axes (Image 10,12) due to their interacting flux (clearly shown as the "hole" in the middle

12.

of image 14). It is proposed that matter itself prevents these two axes from merging into one Ideal Axis (explained later in detail). Otherwise they would become the meridian of the new "ideal" field and fully manifest a new *single* axis perpendicular to the old ones (image 12), forming the Ideal Torus (image 11).

The interior axis is a thin vortex traveling as a ray which defines a spiral wave, while the exterior is a bloated vortex occupying a specific space acting as a particle or field, yet they are both merely One cycle, inseparable from each other, just as is the magnetic field around an electron in a wire. Image 11 and 12 may thus be considered as a model for any individual point (particle) or ray (wave) of any field, or the entire system of any field, because it is a recursive model, happening exactly the same throughout the entire field, albeit in different scales. Just as the divided magnet becomes a smaller aspect of itself.

Notes: 1. Image 12 may be referred to when imbalances in systems keep them from fully unifying. 2. Spin is relative to the perceiver. Looking at image 11 from the north side, all things seem to spin clockwise, however, if the observer perceives the same model from the south, the entire system appears to be spinning counterclockwise.

13.

Notice how images 13 and 14 are similar to the p and d orbitals in image 15. Further note that the "s" orbital can be created by lining magnets up axis TO axis, thus easily creating the Ideal Field (image 11). The closer the physical geometry of the combined magnets is to a sphere the more true to the Ideal Field Model the final field will behave.

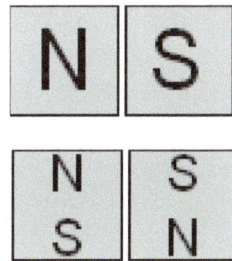

14. Top and Bottom: Ian Smith, www.ian.org

These images are two views of the same configuration, two magnets lined up axis BY axis (as opposed to axis TO axis.) The phantom axis vortex is revealed as the "hole".

It is proposed that the true path of all field lines are spiral in nature, and that the axis is also a vortex, neither of which is easily viewable with the classic methods used in images 6 and 7. With 2d iron filings and a 2d compass one can only plot a 2d vector. This is why the spiral is not noticed in such limited past analysis of magnetic fields. Thus, to observe magnetic fields in 3 dimensions, it is accomplished well by employing Ferro fluid as in image 13 and 14. Ferro fluid is basically iron powder suspended in light oil which easily conforms to the magnetic field in 3 dimensions with notable detail.

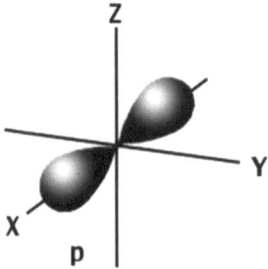

When working with Ferro fluid, at first it is observed that uniform geometries of a base magnet, or uniform charges, simply make a grid-like field, nearly identical with the classic field line interpretations (image 7), and *seemingly* there is no spiral. However, a spiral can be revealed in 2 ways:

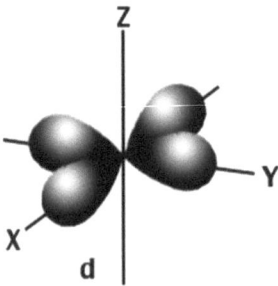

1. By increasing the amount of Ferro fluid on a non-spiraling base form of, a) uniform geometry and b) constant charge. The variable is amount of fluid (matter) which consequently increases the distance of the Ferro fluid's surface from the center of the magnetic flux. (See image 16) At first the Ferro fluid seems to assemble in a uniform grid. As the amount of fluid begins to increase and the "spiked tips" of the Ferro fluid get increasingly further from the center of "charge", the tips of those points will travel in a spiral. Therefore a spiral is traversed by a point from the center of charge to its perimeter.

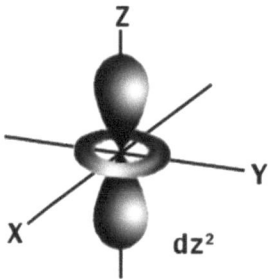

This explains why a static model cannot observe a spiral, because it is only measuring One radius from the center, namely, the final surface. To illustrate: Imagine measuring just one point on the last outer sphere of image 16, instead of considering all 5. It could very well be concluded that the spiral is a 4th dimension of "growth" or motion/time.

2. Simply creating a spiraling base form (image 17) with constant charge, and then adding Ferro fluid, will cause the fluid to move in a spiral. This is unlike the other forms where the fluid merely assembles thereon without visible movement. The variable is again: Ferro fluid amount. This type of base form allows the fluid to seek its equilibrium on the X, Y, and Z-

18

axis unlike the other forms. Note that even though Ferro fluid always forms 3dimensionl models, if the faces of the base form have only 2 planes of movement, the dynamics of the force are not properly modeled because it is limited by the 2-D planes of the base model. In forms where each face is a 2-D plane, the third "spiral" Z axis is created by the fluid itself, thus the spiral motion is occurring within the fluid, below the surface and cannot be observed accept by the first method. This motion is captured clearly by the spiral base model.

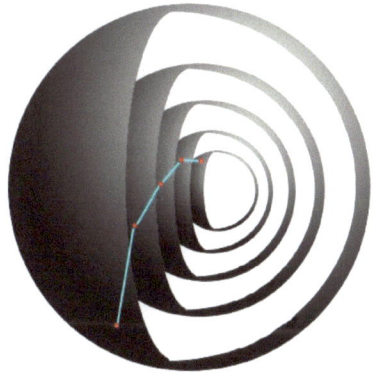

16.

18.

NOTE: for sake of simplicity Image 16 only roughly plotted this spiral on the X and Z-axis, however it would also have deviation along the Y-axis as well. Image 18 depicts the initial doubling diagram (pg 9) described in a DNA spiral, in green is the field between the pairs.

17.

The spiral can further be revealed by the Field Model itself (pg11). Simply by following the pattern of numbers below or above the model's surface will show that the pattern is three-dimensional. Not only does it show the spiral but it also reveals the nature of each number in more detail. For example: cube a) is the cube of the numbers around +8. Notice the +8 is directly in the center of the cube. Next to image a) is the cube of -4. Note that − 4 is always next to +8 in the number pattern (Image 0).

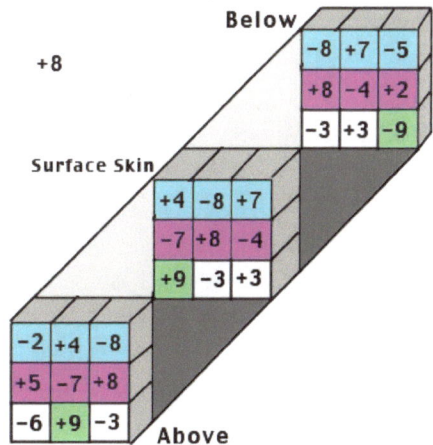

19. a

Now it can be seen how these two numbers interact in more detail. Observe the balance that is maintained. This reveals even more insight to the "signatures" of each number. It is

19

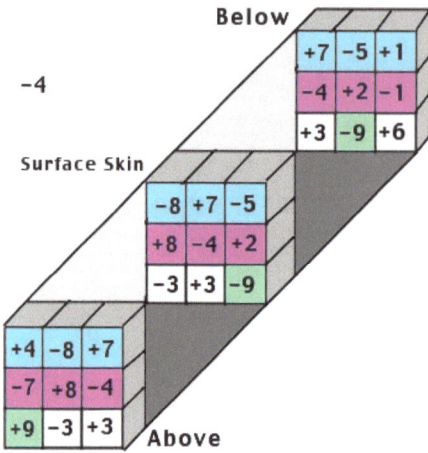

19. b

Below

-4

Surface Skin

Above

Below: +7 -5 +1 / -4 +2 -1 / +3 -9 +6

Surface Skin: -8 +7 -5 / +8 -4 +2 / -3 +3 -9

Above: +4 -8 +7 / -7 +8 -4 / +9 -3 +3

proposed that the study of this pattern will lead to thorough knowledge of the numbers as being 9 aspects of One Force of manifestation. Once each aspect is fully understood, and how it reacts with the others, patterns will then be able to be skillfully assembled with energy itself, resulting in truly man-made materials. From strings, to atoms, to molecules, man will have a new level of manufacturing. The twisted structure of complex proteins will become clear, massive fields around galaxies will be harnessed. The universe will become a sandbox to mans' creativity.

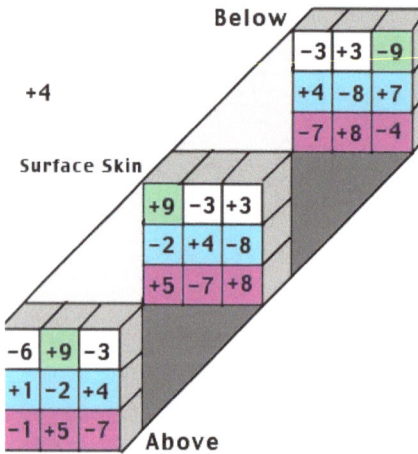

19. c

Below

+4

Surface Skin

Above

Below: -3 +3 -9 / +4 -8 +7 / -7 +8 -4

Surface Skin: +9 -3 +3 / -2 +4 -8 / +5 -7 +8

Above: -6 +9 -3 / +1 -2 +4 / -1 +5 -7

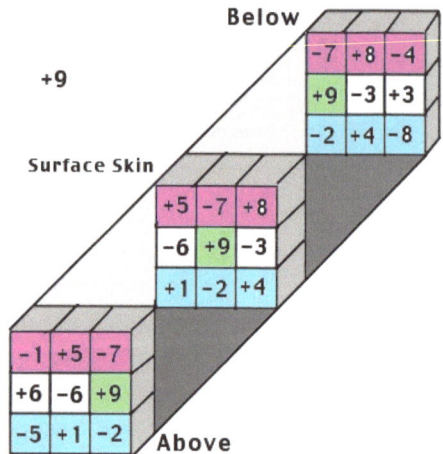

19. d

Below

+9

Surface Skin

Above

Below: -7 +8 -4 / +9 -3 +3 / -2 +4 -8

Surface Skin: +5 -7 +8 / -6 +9 -3 / +1 -2 +4

Above: -1 +5 -7 / +6 -6 +9 / -5 +1 -2

All that any unit of matter is, is a symphony of vortices, an orchestra of these "aspects". A field or particle always attempts to get as close as possible to the Ideal Model, governed by the most perfect geometry throughout all scales of size from: 1. The specific geometry of the sub atomic particles composing the atoms of the material used(electrons, protons, etc). 2. The geometry of the alignment of those atoms making up the material. 3. The geometry of the pieces of said material employed. 4. The geometry of the array of those shapes. Therefore, by careful choice of material (micro-geometry) and its geometry when assembled (macro-geometry), any field can be manipulated and directed.

20

The most perfect and symmetrical geometries allow Force to travel unobstructed. Any impurity taints the force resulting in an imperfect field directly related to that the impurity, this then directly causes the "characteristics" of that particle or field. Thus, carefully crafted imbalances can direct force. For when force is perfectly balanced, it interacts with everything equally and thus appears to be unreactive because of such uniformity. "Precise imbalances" and "perfect imperfections" are key. This force is the so-called "zero point energy". Whisps of which can be seen in permanent magnets due to their small degree of symmetrical internal geometry (the alignment of electrons). It is this force which this paper shall Model.

To conclude this chapter on the relationship of the electromagnetic field to a spiral, and that spiral being related to a system of numbers; here is a small note on how the humble number system may be even more profound than what was previously thought:

Notice in the Image below that the 4 and 7 are mirror images rotated 90 degrees. The 2 and 5 are mirror images rotated 180 degrees. And the 3, 6 and 9 are direct depictions of a vortex, the 3 (shown in grey on right) is a 6 or 9 with the face removed on a line tangent to the inner circle and its own tail. Notice how they hold mirror positions on the page 9 diagram.

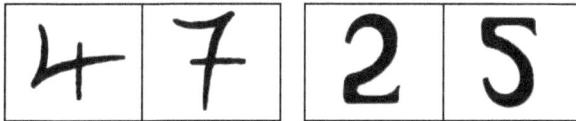

Those of Christian faith may note the "666" similarities, does this make the vortex evil? The author thinks not, unless all of creation is evil, for it is formed by the force of the vortex. But maybe this world of matter *is* evil as the Gnostics have long said. So then what is "evil" in this sense? Imbalanced, impure, imperfect energy perhaps?

As for the 1 and the 8: The 8 is the Infinitely Curved, The Mobius Strip. While the 1 is the Infinitely Straight, The Ray. Both exact opposites of each other. This draws an interesting connection between 1 and infinity, showing that infinity is the mirror image of ONE. This may be the root message in Deuteronomy 6:4-9, as well as many other beliefs which equate infinity to 1.

The above images suggest that a toroidal galaxy is a calculable system based upon recursive mathematics. The simple Julia set fractal on the left is an exact recursive formula, and also "Ideal". Just as with the Ideal Gas Law, No Gas behaves Ideally. Nature is far more complex, the law simply helps humankind work with and understand such complexity. The same is true for the Ideal Field Model. Gas, Plasma and Fluid mechanics all propagate in complex recursive iterations, which are all more clearly grasped with the Ideal Field Model.

2. Universal Mechanics

As shown in the previous chapter, the model can be used for any particle or field; it shall therefore be used to briefly describe the Ultimate Field, that of existence itself, and to use that description as a foundation for a more detailed discussions on the applications of the model. Note: There *seems* to be some contradictions in the following paragraphs which do not become clear until the section on time. The below is described in terms of *linear time* for sake of simplicity, however there truly is no such time, as it is merely a phenomena resulting from the limitations of human consciousness, which will be addressed later.

The Universe is the set of All possible vibrations which vibrate in harmony creating one perfect field of one solid tone – Silence. It is a sea of perfectly organized energy in perfect balance and vibrating at this perfect frequency. All that is, is contained within this field and composed of this energy. All things that exist are localized "imbalances" of this perfect field, each being a vortex itself, similar to the 9's on the Ideal Field Model.

It is proposed that the torus model is a two-way vortex in this "Field of Silence". The center of the torus being simultaneously the entrance and exit of energy from this field of infinite potential, recursively mimicking the model to an infinitely small scale. The model is therefore of an anomaly in the field, one end ejecting energy in a vibration relative to the anomaly's properties, and the other end likewise harmonizing energy, returning it to Silence.

When the Field of Silence (the Universe) is "out of balance" at any point, its frequency of vibration changes, this vibration, by its very nature cannot exist in the field of Silence and it is ejected. However, since that field is the only field that exists it ejects the energy to a specific part of itself which vibrates in harmony with some other aspect of itself, thus maintaining silence. However, a perceiver observing only that one point would observe "sound" since the observer is not capable of observing the entire Universe at the exact same time, else the perception would always be Silence. This "sound" simply means: "any vibration other than silence". Therefore, All things come from this field, for all things are vibration.

These "vibrations" are the "friction" of the motion of the One Force flowing through the vortex creating a specific frequency or "note" or "chord". Each vortex having specific properties based on the 9 numbers; the strength of these properties is based on the relative "size" (amplitude) of the vortex and the symmetrical perfection thereof. These properties interact with each other as shown in Image 19. The various vortices when interacting with each other literally create a "symphony". A symphony is the measure of the compounded vibration of all units in a group. Therefore, *All things* are, to more or less complex degrees, symphonies of vortices. Furthermore, this union of vibrations forms a collective vortex as in Image 12, uniting the group as one. The more perfect this assembly, the closer to Image 11 this collective phantom vortex becomes.

Think of a magnet as a balanced anomaly, a gate that cannot grow larger or smaller, therefore a flux continually circulates though it in the exact same proportions, it is thus a specific "note". A magnet is our torus, as the early images show. Thus the constituents of the atom are also magnets, which are tori, which are vortices, which are "notes" which are anomalies in silence. Thus, String Theory may merely be attempting to describe very tiny toroids, whose vibrations are notes in the Titanic One Song (uni – One; verse – Song), the Universe.

What then creates these anomalies? It is simply geometry. As previously stated, the universe is perfectly organized which allows it to have a perfect vibration. No vibration can be efficiently transmitted through a medium (which is itself a vibration) that is not symmetrical. Therefore symmetry is directly proportionate to transmission efficiency. The highest degree of symmetry is the ultimate balance of the Universe as a whole. Therefore energy flows though it perfectly, yet so perfectly that nothing can be differentiated. Any less symmetrical material, even the most advanced and symmetrical material made by man, will lack some degree of efficiency. Not to mention the fact that the material would have to evenly cover the entire universe to achieve ultimate perfection, thus some "friction" will be present in all works less than the perfect "One". This most symmetrical form is expressed by the Ideal field Model as it cycles through itself in infinite recursion without deviation on any level of magnitude.

Whenever sufficient symmetry is present in some part of the universe it momentarily mimics the architecture of Silence, as a recursive harmonic reflection and thus "The Energy" flows through it with an efficiency directly related to how closely it has mimicked perfection. This "conduit" (vortex) then affects the surrounding environment (which is simply a panorama of vibration) according to the 9 properties of The Ideal Field model.

As an illustration: Say a part of the universe, on a microscopic scale, for whatever reason, natural or man-made, happens to mimic a part of silence. Suddenly an infinitely tiny vortex manifests (a string?) allowing energy to flow forth from silence which effects surrounding vibrations accordingly, possibly even causing more symmetrical geometries, in the environment, thereby initiating other vortices. These other vortices may be in the right position to interact with each other. The flow of energy from the first vortex may be balanced enough to fortify its structure, so that when it comes into contact with another, they do not meld together. Instead, they form a dynamic interaction (such as image 12). This is the origin of the compound particle (such as an atom). This particle then, also acting as a vortex, continues the same progression of interaction, ultimately forming the entire material universe.

24

It is the nature of this force to thus generate complex systems with each of their individual parts, from atoms to galaxies, all attempting to mimic perfection, striving to grow into a more symmetrical form. All of this happens by the same simple recursive mechanism of the Ideal Field Model compounding infinite pieces into infinite variety. The Ideal Field Model is thus the Universe whose infinite pieces act exactly as itself. Thus all matter are vortices, which are simply the initial energy "bound" into self-similar feedback loops.

Before continuing to the next chapter, here is a brief summary of the key points thus far:

0. The Universe

0. First, there is nothing but One Sea of the One and Only Energy in perfect balance thus, indistinguishable from any other aspect of itself, it therefore seems to be a Silent Void. ALL is made of this energy. (For this document this energy from hereon will be called Zero Point Energy or "ZP")

1. The Totality of #0 is simultaneously at rest and silent, as well as in dynamic movement and incredibly loud, emitting every possible sound. The human mind cannot fully fathom this, and ultimately only perceives limited parts of the system, never able to perceive the totality and comprehend how such noise balances into Silence. Therefore it appears as if there should be a "beginning", even though it is a birthless and deathless system beyond the realm of time. Time is a phenomenon of the limited human consciousness, which analyzes each independent space/time event with varying degrees of intensity in specific sequences. However, all events already preexist in the Silence, otherwise it would not be One and only One, for infinity is equal to one because there is only One Infinity.

2. Assuming #0 and #1 are true, even though all phenomena have always existed, being limited humans, the mechanics must be analyzed from this limited point of view of linear sequences. Therefore 2 worlds must be addressed:

A. The world of #0, which is a perfect Sea of the Purest Energy, the One Energy, the "Zero Point" energy, or the Luminiferous Aether. This world is beyond the bounds of human comprehension in its titanic vastness; and is Perfect Silence.

B. The World of Phenomena, or World of Matter. This is the limited view of things perceived by the limited human consciousness, which tends to see things as "separate", when in all truth "separateness" is an impossibility (else One does not have One Unit as shown in #1). This world is the "noise" perceived when unable to perceive the whole. Thus, time exists here - which is a direct result of separation. For to perceive separation, things must be compared to other things which then gives rise to "sequence". Therefore, all phenomena in this paper shall be discussed in sequences and in relation to other "separate" phenomena.

I. The Basic Linear Cycle

0. Being that there is no beginning and no end, just One perpetual cycle, the cause of any "point" is therefore, paradoxically, preexisting stress(es). Thus it can be true that any manifestation, could be the very "first" event ever, as the following sequence is a perpetually recursive cycle. No different than the "first" link in a circular chain.

1. Localized activity in the World of Phenomena creates conditions favorable for the formation of a vortex, possibly incredibly small (possibly called by some "strings" or by others "gravitons").

2. The specific nature of this vortex interacts with the specific nature of preexisting forces (other vortices) catalyzing new events, and eventually leading again to step 1. A vortex may grow by attracting and/or combining with other vortices, or be "dissolved" by them. (Both explained in detail later) Therefore, all things are made of "bound" Zero Point energy, whose mechanics follow the Ideal Field Model.

II. The Basic Forces of the Vortex

0. A vortex is simply ZP energy which has become intertwined with itself by mimicking the Silence on a lesser order of magnitude, thus it is "bound" to itself in this specific form, which is the Ideal Field Model (image 11). It has become a "self-interacting" feedback loop.

1. The formation of the initial vortex is aligned to a plane depending on the surrounding forces. Its alignment may be in constant, violent fluctuation, or quite even and stable; all dependent on its specific nature and the local environmental forces, both closely related (shown later). However, it tends toward organization and stability, as discussed in the following section.

2. These vortices interact with each other depending upon their "Texture" which is the 3D map of their "surface" generated by the frequency, amplitude, wavelength etc. of their specific pattern of "self-interaction" (vibration).

3. The Mechanics of Motion

So far we have examined the torus as the Ideal Model, the Spiral nature of electromagnetic waves, and a view of the Universe as an Ideal Field. Yet all of these discussions were predominantly still and unmoving, therefore Motion must now be discussed.

There is Only Motion. Motion gives rise to All phenomena from magnetism to strong forces, simply by recursive and cumulative interactions. The first thing to move was the Silence. In complete stillness, nothing perceivable exists because perception can only be had by relating one movement to another. This first (and only) energy, ZP energy, is so fine that it can flow through any "particle" that it creates. Even though it is not divisible into any known unit, for simplicity's sake, a "unit" of this pure energy, smaller than anything known, will be imagined.

Now imagine the totality of Stillness in total balance, then imagine it has one "unit" that has moved. The instant one thing moves in a system, all things must move however slightly. Thus the entire universe is moving as a consequence of this initial unit. This unit is therefore "The Origin", or "The Prime Mover". (What caused it to move is irrelevant because there was never such a beginning, only the limits of the human mind cause it to see such linear restrictions, thus Only for Example is such a "beginning" presented).

This movement causes an instant and total chain reaction, because All of the units are touching each other. (Imagine a jar of tiny, soft rubber spheres, it is their ability to flex that allows for motion and disallows a void to ever form. For a jar of hard marbles, unable to flex, would be an immovable crystalline solid.) When moved, the "unit" pushes on units ahead of it causing an energy density gain in the direction traveled. This transfers part or all of its "movement"(energy) to all touching units, which in turn likewise transfer it to all other units in the system. This cycle continues until the space where the initial unit first occupied is filled instantly by the last particle moved, never allowing a void. However, the last particle moved restarts the entire cycle over again because of the "potential void" which its movement has now created. Thus the potential void is infinitely being filled and recreated by movement itself. Movement is thus a cycle of propagating Energy Density fluctuations within a medium.

This propagating energy density cycle causes a spiral vortex as seen at any drain, which results from its composing units moving towards the drain (the potential void) yet impeded by units ahead of them which slightly deflect their straight path into a spiral path. However, this happens in 3-Dimensions unlike a 2-D drain. The initial movement erupts a spiraling energy density fluctuation from the center in the direction traveled which cascades in a spiral back to the center potential void. (See image 11, and realize that the Axis *is* the direction of movement.) Thus every particle *is* an Energy-Density Fluctuation, which is a vortex, *and when moved* also causes a vortex in its carrying medium directly proportionate to the force of its accumulated movement or "energy density", (like the moving charged particle and its surrounding spiral magnetic field). All instances clearly obeying a recursive application of the Ideal Field Model.

When such a "motion cycle" is well balanced in a portion of a carrying medium, its energy does not dissipate and the cycle is constant. This is the origin of "bound" ZP Energy (matter), a vortex composed of multiple units which act as One unit, they are literally the force of movement trapped in a repetitive cycle. These "particles", or "concretions" of a medium, now distort the flow of "unbound" or "free" energy around them proportionate to the specific pattern (geometry) of their individual energy density fluctuation cycle, which creates their exterior "texture". This disturbance may in turn spawn other vortices as eddy currents, which may also balance and become "particles" themselves, repeating the cycle.

When enough particles of a similar nature are in close quarters, they behave as a "medium", and the entire motion-cycle repeats itself on this higher level of magnitude with these larger "units" now acting as the foundation units. Again applying the Ideal Field Model in a recursive manner. In this way particles and corresponding mediums, become increasingly more complex, and soon it is better to call them "systems" as opposed to "units" or "particles".

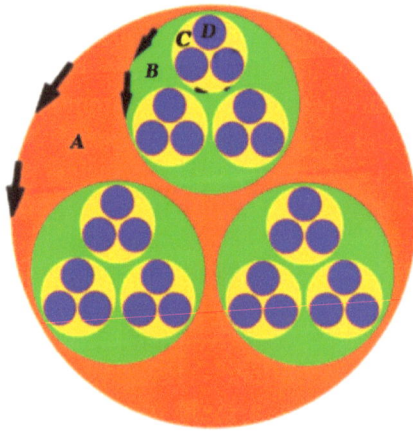

20.

To illustrate, observe image 20: Imagine "A" as the fine ZP Energy units composing all things; B as larger units like strings; C as even larger units like the electron, and D as even larger units like atoms. A creates B creates C creates D etc. Each new system acts as an Ideal Field Cycle in smaller and smaller magnitudes. The entire image is the first and largest Ideal Field Cycle, thus all things within it are its "energy density fluctuations" propagating from the Prime mover back to its Potential Void (which are practically one in the same). One can now see the Universe as a sea of recursive Energy-density-fluctuations caused by the motion of The Prime Movement, to which all things will return. (Note: The shape of image 20 has NO significance, it is merely a simple example of recursion)

For further detail: All bodies within the D system would operate under the influence of the forces of D, C, B and A. While system C's bodies would operate under C, B and A; and likewise, system B's bodies would operate under only B and A. While *All* things would operate under A. It is in this way that all things (ZP energy) are always traveling towards the Initial Potential Void, even though they may get caught up as "bound" energy in countless minor cycles (forms) for a time.

Important Note on Energy-Density:

This paper deals primarily with "Energy-Density", not to be confused with common material density. $E=mc^2$ shows that matter *is* energy, thus material density is a measure of energy. However, energy-density is the sum of this material density *and* all other acting energy such as charge, temperature, spin, etc. Therefore, a stationary object's energy-density increases as it is set into motion because of the kinetic energy transferred to it. The same is true if an object is heated, its *material* density may lessen as it expands, but its total density has possibly increased (depending on the degree of expansion). To illustrate: picture a stationary diamond, then a spinning diamond, then a heated spinning diamond, each having progressively more total energy per unit space.

To illustrate further: Imagine all of the circles of image 20 spinning counter clockwise and also spinning all of the circles that they contain. Being that Motion is transmittable and cumulative, the forces in one system cannot be fully understood without knowing the forces of the containing system (the medium). For example: If the spin of D was measured and then D stopped spinning, it would still continue to move on the parameters of C, B, and A. So by measuring only the spin of D, one would not have accounted for all of the energy within that particle or system. By neglecting this, one would end up with a great deal of missing energy. Dark Matter Perhaps?

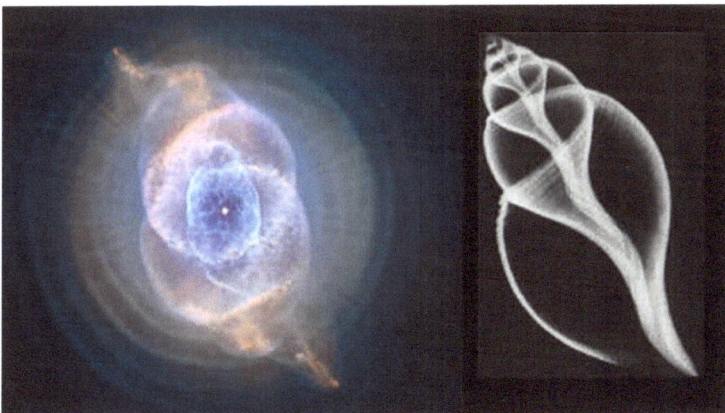

Left: nasa.gov   Right: Orphaned

The Cat's Eye Nebula shown to be a dual spiral vortex with striking similarities with the spiral of a common snail shell.

4. Texture

The last section equated the basic recursive cycle of motion to the Ideal Field Model and briefly mentioned a "Texture" on the surface of energy which has become bound in a self-interacting feedback loop (matter). This texture is the pattern of the motion itself, and can easily be seen in a glass of water; when the glass is vibrated, the motion (vibration) causes a texture change on the water's surface. To understand the mechanics of motion in greater detail, this "texture" will now be examined.

As was explained in the previous section, motion through a medium causes that medium to compress together, and thus gain in energy-density at the point of contact in the act of transferring the motion (energy). This happens in a specific pattern, ideally as the Ideal Field Model, but more often in inferior patterns, which give rise to the specific characteristics of matter. Occasionally portions of the medium get "bound" into this higher energy state and become a recursive reflection, on a lesser magnitude, of the movement pattern which has spawned them; thus becoming a "particle". Therefore any particle is "denser" than the medium that compressed to create it, and possesses a unique pattern of vibration.

The Ideal Field Model is thus an energy-density gradient with the center as the most dense, to the exterior as the most rarified. The cycle of energy-density propagation (follow the red arrows on image 21) begins at the center of the Model as one compact force of the initial Prime Movement. It then spreads out as the interior vortex in the direction of the axis vector rarifying toward the surface. Then it cascades in a spiral on the exterior surface to the opposite pole and re-compacts on its return to the center to fill the potential void. Remember that the "9's" on the model (represented as the small blue circles on the violet band) are also minor vortices. They too follow the same density gradient with the center of the vortex the least dense (like the tube of air that runs through the center of a water vortex). This center column of the vortex, though less dense, also follows the gradient, getting denser toward the center of the model. The exact nature of the final form created by this cycle is dependent upon the characteristics of the specific wave traversing the cycle.

21.

22.

23.

24.

Image 22 shows the energy-density recursion of the Model. In reality, there would be infinite levels of recursion so close together that they would be nearly indistinguishable, however to exaggerate this, only three levels are shown in the model for clarity, a blue sphere inside a yellow sphere inside a violet sphere. The blue sphere being the most dense and the violet sphere the least dense. Their collective axis vortices get smaller from the violet to the blue which causes the smooth interior vortex as seen in image 21, the same vortex seen when draining water.

Image 23 is a hypothetical energy density cross-section of the solid Ideal Field. Density is denoted as darkest being the *least* dense. (The gradients are exaggerated for clarity, normally boundaries would flow smoothly one into another.) It is this network of densities that creates an "architecture" in a system. Image 24 highlights in red this "architecture" surrounding the vortices, which is the result of the "boundaries" of their flow, the highest densities being where the edges of vortices meet.

Image 25 shows how this gradient would basically appear on the exterior surface of a system. Image 26 roughly details the flow of energy of the minor vortices while still maintaining the spiral path of the Major Pole vortices. (These spirals must be imagined in 3-dimensions, each descending in towards the center of the system and then out again).

Note: do not allow the limitations of these examples to restrict your visualization. The possible number and size of vortices could have infinite combinations, the examples shown here are simply for clarity. It is also important to understand that the "architecture" and the "vortices" are **not separate!** They are simply easier ways to visualize a *continuous* **energy density gradient**. For example: Image 27 shows that the earth and its corresponding magnetic field are at different magnitudes of energy-density from the center outwards. Even the magnetic field rarifies further from the center. The earth and its field are too often looked at as "separate", when they are actually One System. However, they can be analyzed independently due to the recursion of the model (image 22). To further illustrate: The electron "cloud" and "nucleus" of an atom are likewise simply levels of magnitude of energy-density on a continuous gradient. Likewise, the nucleus of the atom is also a continuous energy-density gradient with merely the potential to be split into those units called "protons and neutrons". Lastly, the visible part of the Sun is only a tiny part of the titanic "solar system", whose most rarified levels extend beyond Pluto.

The density of a system decreases from its center to the perimeter, and all of its composing units have a proper level of density within it. Any unit within the system will be a distance from the center directly dependent upon its own energy-density. All units are forced toward their proper layer of density by the action of more energy-dense units.

The exterior texture of a system is the result ,of its energy-density fluctuation pattern, which is a spiral wave cycling over itself along a spiral path, forming a net "disturbance" on the surface from its constructive and destructive interference with

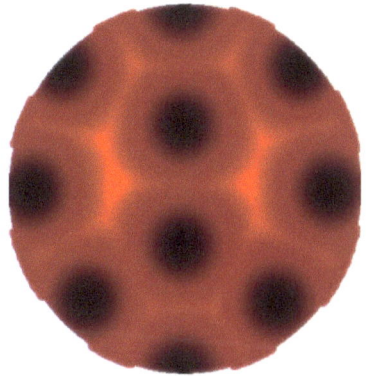

25.

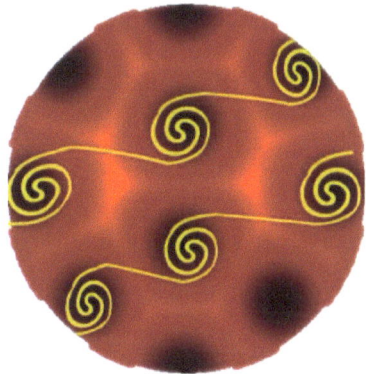

26.

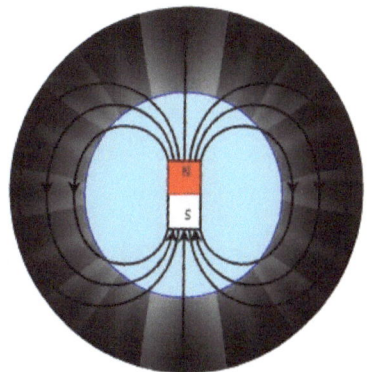

27.

itself. Understand that with infinite recursive levels of magnitude, the "surface" is defined by the "level of magnitude" one is examining at a specified distance from the center of the system. This disturbance can be

33

28. a

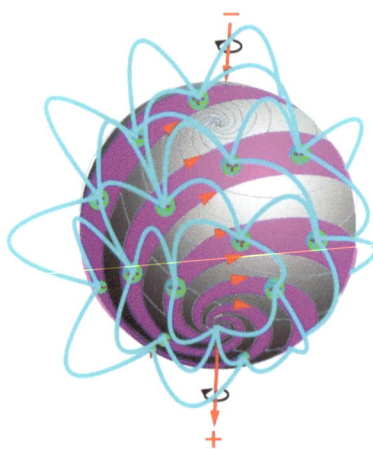

28. b

28. c

violent, uniform, irregular, etc. all depending upon the nature of the medium and the characteristics of the cycling wave.

The minor vortices in a system are the "Valleys" of the net waveform whose cycle is the complete circuit between the two major pole vortices. Their flux (the dynamic wave pattern) spouts like a spiral fountain out of the positive vortices and into the negative ones (image 28a, b). In a "disturbed surface" the more dense energies, closer to the center, occasionally are energized by the flux flowing to higher magnitudes and thus gain enough energy to temporarily rise up from their normal "rest" level of density into a more rarified level. The "field lines" they follow may not be the spiral of the Ideal Field due to the medium's limitations in acting as more Ideal energy, coupled with its energy requirements to do so. The more "Ideal" (pure energy) the medium being moved is, the closer it will obey the Ideal Field Model.

This is illustrated with a low amplitude vibration in a glass of water. First the surface texture will hardly change, but raise the amplitude to a significant level, and the water may actually splash out of the glass from the violent surface fluctuations. This can be further illustrated at the base of any water spout (image 29), the air vortex represents the more rarified vortex flux and the toroidal mound at the base of the spout is the denser medium, in this case water, being temporarily raised into a more rarified order of density, while attempting to mimic the vortex dynamics on a higher level of magnitude, yet with limited energy. One can observe the same spouting toroidal mounds as Solar Flares on the surface of the sun (image 28c).

An "Order of Magnitude" of a system is simply a smooth sphere at any distance away from the center of the system, the larger the distance, the larger the order of magnitude. Image 30 shows this in violet and red, red being the highest order of magnitude of the system. In contrast, green shows a single level of energy-density by highlighting

29.

one specific shade. Notice how an "order of magnitude" (violet) will extend over multiple energy-densities, and one energy-density (green) will traverse multiple orders of magnitude. The one has a fine texture of numerous densities, and the other a rough texture of one density. Imagine the violet ring as an "electron cloud"; the denser areas would be the "electrons".

It is these "textures" which govern the interactions of systems, because the differences in energy-density govern energy flow in the Universe: Basically, less dense energy always flows *around* more dense energy unless it is forced into a point of obstruction (explained later). Less dense energy can be made into more dense energy by imparting motion (ZP energy) to it with pressure. Motion is transferred when two systems touch, in proportion to their pressure.

30.

A simple way to visualize this interaction of systems is by gears, because this texture is "toothed" in a sense. This makes it easier to visualize and map the complex interactions of frequencies and amplitudes stirring around inside a system or between two systems. These "teeth" are a general map of a system's potential to engage and effect another system in proximity, from solar systems to atoms, based off of its cycling waveform(s). Frequency and amplitude are the number of gear teeth and size of those teeth respectively. Image 31, in red: 1. would be amplitude, and 2. would be wavelength, (the adjacent image represents a compound waveform).

Images: 28a. vario-fountain.com 28c. nasa.gov 29. noaa.gov

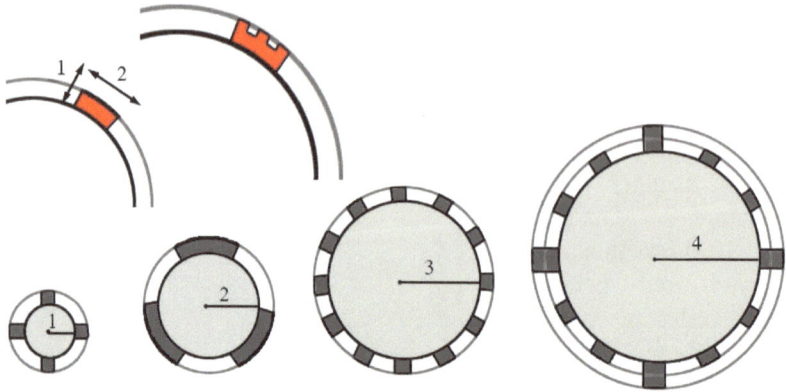

31.

These "gears" represent the pattern of a system's waveform. The "peaks" of the wave have precise domains, which can be thought of as "teeth". The constant internal fluctuation of a system causes the peaks to cycle across the surface resembling "spin". When energy is added to the system this spin increases. These "teeth" represent the system's attributes when engaging another system, and so too do the "anti teeth", which would be the valleys of the wave.

For example, the above gear of r4, may interact with r3 temporarily in cycles, or with a smaller "gear" temporarily engaging between the longer teeth, only to then get violently struck away by the longer tooth when it cycles through. However if the placement of the smaller gear was such that it only interacted with the outermost Order of Magnitude, a continuous cycle could be maintained. These examples are all very simple, and when compound waveforms are introduced, the texture of the "gear" and its cycle could be very intricate as well as its interactions with other systems. This "texture" of any order of magnitude of a system is composed of only the waves of a given energy density that emanate *beyond* the chosen distance from the center.

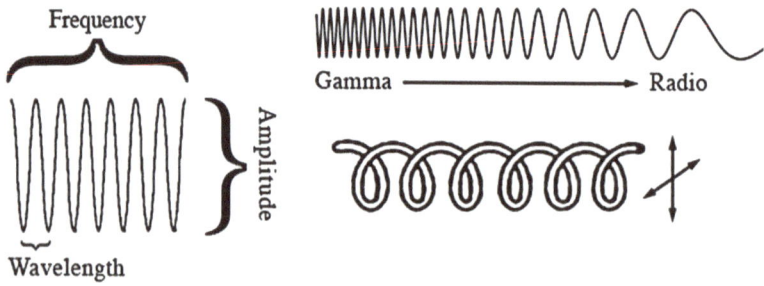

32.

Images: 33. "Elegant Universe" still by NOVA, pbs.org, 34. Gregory F. Maxwell, wikipedia 35. "Floriang" wikipedia 36. MANOJTV, wikipedia

NOTE: Image 35 is the "quintic string", which is physics' present attempt at visualizing the string, notice how it is evolving from the earlier 2-D toroidal form.

36

However, the *ideal* "gears" are 3-D spheres not 2-D discs. If one takes the commonly visualized 2-D wave in image 32 and realizes that it is actually a 3-D spiral, and then closes it into a circle, one gets the early ideas of what a string may look like. (Image 33) However, this is far from the Ideal Model, because the path of the spiral wave is still on a 2-D plane. When the path is made into a 3-D vortex, and cycled *through* itself, the Ideal Field is created.

To visualize these teeth in 3 dimensions, observe the spikes of the Ferro fluid (image 34) as gear teeth, but imagine them not as iron, but as a charged elastomer, pliable yet firm. In the presence of other "gears" such pliability and charge allows them to line up and mesh perfectly. Another example of the texture of the Ideal Field is, oddly enough, a pineapple. (Image 36) Imagine the "teeth" of the pineapple as being different sizes, in certain patterns, and possibly even fluctuating due to internal constructive and destructive interference. Also, imagine the peaks and valleys each as a miniature

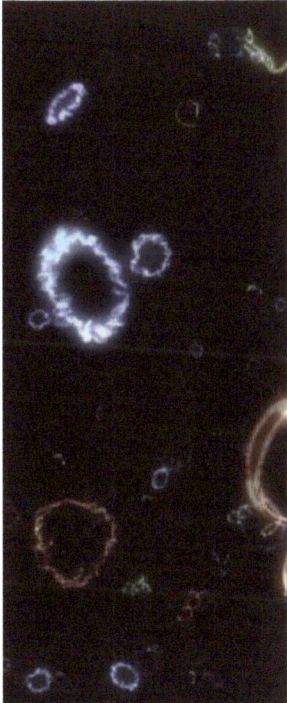

33.

textured vortex as the 9's on the Ideal Model. Just as the vortex draws in energy from one side and spews it out the other along its axis, the pineapple sucks up water from its base and spews out a crop of green leaves from the top.

34.

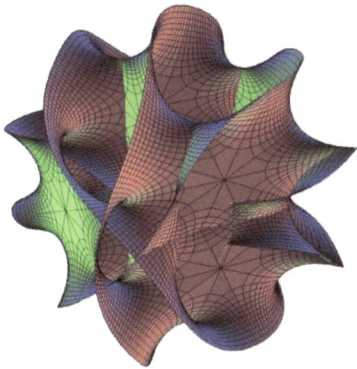

35.

36.

5. Rules of Engagement

To summarize thus far, an initial motion causes an energy-density fluctuation pattern that propagates by the recursive spiral dynamics of the Ideal Field Model. All units of a medium caught up in this cycle are now considered a system or "particle". This system interacts with other systems according to the geometry of its "Levels of Energy-Density", (teeth) which are the direct result of the pattern of the wave and nature of the medium. The mechanics of interaction of such systems will now be discussed in detail:

On the edge of any system will be "teeth" to some degree, even if it is on the Nano-level. These teeth are the peaks of the composing waveform as already discussed. It is common knowledge that when two like waves come together they cancel, but let this phenomenon be modeled in a slightly different manner.

When two waves meet in a spherical wave system, only a portion of their waves can cross. Returning to the gear analogy in image 37 to the right, it can be seen that where two gears mesh, it is the same as two waves canceling. A portion of the two systems literally becomes One, as the peaks of one wave transfer energy into the valleys of the other. It is in this way that similar waveforms can merge, independent of energy-density. This is the action of clouds over terrain, picking up charge by these types of energy transfers, only to return it in a lightning strike.

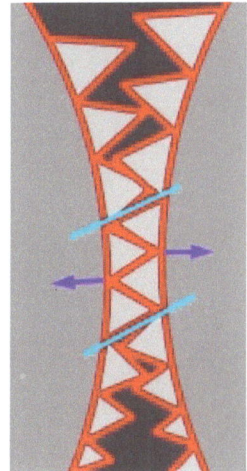

The action of the spiral waves getting increasingly closer to canceling (unity) from the top to the centre and then leaving that unity from the center downwards,

37.

causes the same vortex forces of the Ideal Field's propagation pattern perpendicular to the two axes of the engaging systems, and thus gives rise to a "phantom field" as in image 12, clearly showing the mechanics of the internal dual vortex of image 21.

However, this process gets much more sophisticated due to the various levels of energy-density of a system. For example on earth, the ionosphere is high above, and below that is an atmosphere of denser air, the waters of the ocean, then rock, and then the super heated, super dense liquid metal of the center. Yet despite these various levels of energy-density of a system, every system has an average overall energy density that determines its location within a larger system.

With very tiny systems (particles) within a very large system the interactions are almost unnoticeable, for they easily form "bands of density". When particles get bigger in relation to the system their effects become more pronounced, such as a meteorite descending through all of the previously mentioned levels of earth's system, merely attempting to reach its own proper density level.

But when two systems are of equal size they behave nearly as a particle would, for they cannot fully enter into a density level of the other system for they are too large. Yet the same principles apply and the systems will attempt to enter into each other to their proper level with each similar level of their systems merging to equalization.

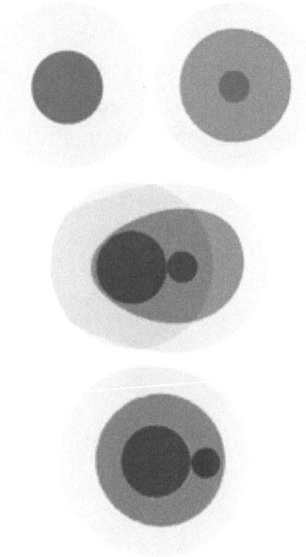

38.

To illustrate: Image 38 represents a basic same size system mergence (not collision). Not only do the systems have the same radius, but imagine that they have the same *total* energy-density. They could therefore exist at rest within the same energy-density level of a larger system.

When these two systems come into close contact, so that their outer vibrations begin to interact (teeth), if the phasing of their waveform cancels (meshes) then that connecting area acts as a conduit for the flow of energy back and forth between the two systems, only at that energy-density level, and just like a merging soap bubble, the two energy-density levels flow into each other and equalize into One wave cycle creating a common medium in which the remainder of both systems then reside.

The force of this flow brings the center of the two systems closer together and with the now shared outer level, possibly adds more pressure on one system, increasing the energy of its waveform and consequently altering its shape and phasing, while relieving pressure on the other, which alters its shape also. Both new shapes will have the same surrounding field assisting the changing waveforms of the next energy-density level to assume a similar shape. Thus the next levels of energy-density will be contacting with extra force and similar geometry to assist a merging. This cycle continues through each energy level until two like energy levels refuse to align in phase (which may even cause a repulsion).

Assuming that no repulsion occurs the two systems could be said to have equalized. The final system, ideally, if it remained the same volume, would be twice as energy-dense. If it remained at the same total energy-density, then it would have twice the volume. Therefore, either change drastically enhances its ability to accumulate other particles, giving rise to the possibility that gravity is merely this simple recursive process over many orders of magnitude.

40

Imagine the same scenario with a meteorite falling though the atmosphere. Any energy-density levels it possesses will be stripped off as it passes through that corresponding level of energy-density, until it rests at an energy-density level equal to its core. Even if it hits the surface and stops, over millions of years it may sink into the earth if its energy-density level was sufficient. (The opposite of this phenomenon

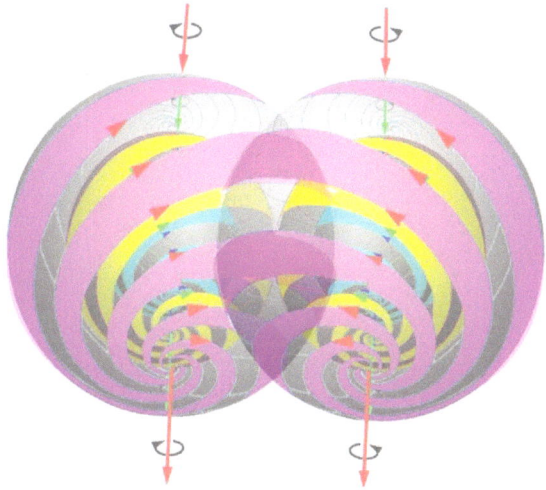

39. (Depicts mergence in 3-Dimensions)

can be observed when burying tires, over many years, they will eventually rise to the surface.) Such is the same for rarified light "rising" out of the center of suns.

The two density levels that have equalized will be two cycles side by side, held close together by their surrounding shared wave cycles, this creates the dynamics of image 12. They literally act as two gears, and their surrounding shared waves will cycle themselves *perpendicular* to the unmerged density levels attempting to mimic the Ideal Field.

However, when two systems in the same density level are *not* close enough to react directly, they may react indirectly in a similar manner. The proper spacing

40.

of two systems "spinning" the *same* direction in a medium causes a phantom vortex in the medium between them that spins the *opposite* direction (image 40). This "phantom gear" is an eddy current. In this manner systems can be attracted by the "current" of another system which would normally repel it. In this case, the system is not attracted to the other system itself, but to the geometry of its disturbance in the medium (phantom gear), interacting with it either fully or partially. However if the systems are too close there will be destructive interference which will cause the systems to instead <u>rotate around each other as one system</u>; in the same direction that they were initially traveling in, in this case, clockwise. Thus same spin systems create a

41

phantom field axis *parallel* to their own, resembling image 11 instead of image 12. This can be observed as the dual vortex atop the planet Venus.

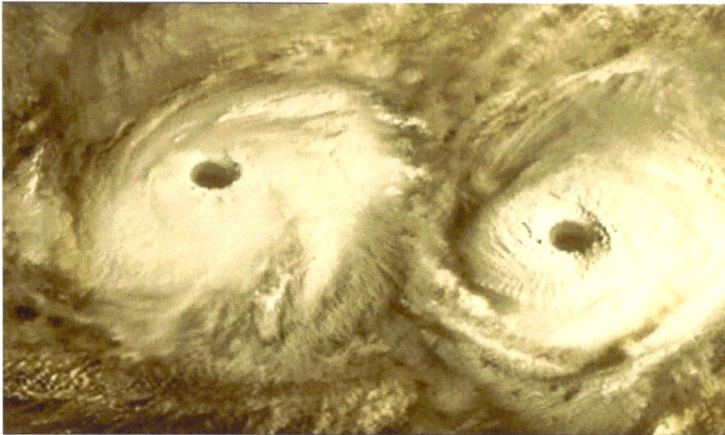

ESA - AOES Medialab

As stated previously, all systems will have ZP energy as their outermost (least dense) order of magnitude, allowing *all* systems to slightly effect, and be affected by, any other system. The reason why ZP energy is the most rarified level of any system, and why density propagates to the center, are explained more clearly by examining the recursion of the Ideal Field Model in greater detail.

Merged energy-density levels allow energy of that level to flow freely through that level unobstructed. When two energy-density levels refuse to phase unify, they "obstruct" each other. When a system is moving and comes into contact with such an obstruction, force (energy) is applied to the obstruction in the direction of motion (image 41). This energy gain in the obstruction causes its waveforms to distort, however slightly, depending

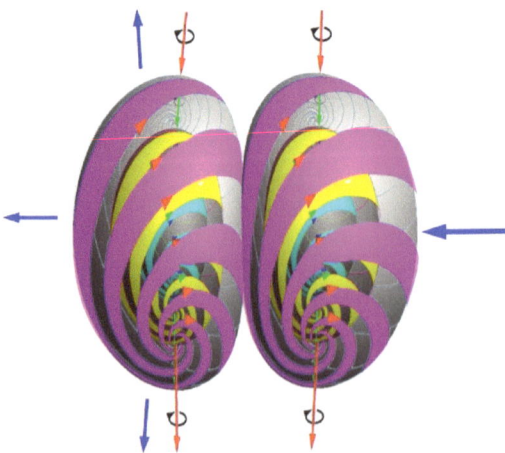

41.

upon the energy-density of the surrounding medium and the applied force. This wave distortion transfers to the other side of the system by the cycle of its waveform. Now distorted and with higher energy, it interacts with the systems it was already in contact with, such as the carrying medium itself, and transfers some or all of the force to them. These systems do likewise, reacting to the higher energy, and thus, the

42

energy-density propagation cycle is born out of an initial concentrated force which diminishes as it is spreads to more and more systems. The initial transference of energy happens because ZP energy is infinitely small, so it will always be able to resonate in phase with some part of the obstructing waveform.

This cycle is a fractional recursion of the initiating force spiraling out in a vortex from a compact center, rarifying as it progresses. Thus motion itself moves in fractals whose pattern is the signature of the initiating force through the operating medium. The more pure the medium, the closer to the Ideal Field dynamics it will behave.

Thus in all systems, as the Ideal Model dictates, the most concentrated energy is the center initiating force. It is this force that generates every subsequent energy-density level by spreading itself out over them. Thus, Ideally, the center most particle's *total* energy is the same as its outermost perimeter, as well as any level in between, forming a continuous gradient. However, the volume of the center is infinitely small and the volume of the perimeter is infinitely large so the energy-densities are opposites. The recursion of this cycle of spreading out and re-compacting to the center is why all systems' density propagates toward the center, causing their highest density to be in the center.

This is reflected even in the atom. The proton and the electron both have the same overall energy of 1.6×10^{-19} J, however the proton is 1836 times more massive, making the electron 1836 times more energy dense. Yet, when in a system together the tiny electron now circles the proton in a cloud 10,000 times the volume of the proton repeating the recursion by effectively diffusing its energy density to 1/10,000 which is about 1/5 that of the proton's energy density.

These opposites can be further examined as a charged particle and its magnetic field. A moving charged particle causes a magnetic field, just as the initial Prime Movement of the Ideal Field Model causes the spiral cascade along its perimeter back to the center. Thus the charged particle is the initiating force traveling on its vector, which is the axis of the spiral magnetic field that it creates in recursive proportion to it own force, obeying the Ideal Field Model. Similarly, every point on a "field line" could be considered as a tiny particle with that field line as its axis and vector of travel.

The Ideal Field is infinite in this constant recursion. Every part contains an exact image of the whole. Thus it must now be understood that any collection of a medium in a self interacting cycle is a "particle", and any portion of this "cycling medium" that gets into its own independent cycle is a new particle on a lesser magnitude. Therefore all things are simultaneously particles *and* mediums, for a newly created "particle" is a particle to the medium that it was spawned from, but it is a medium for particles forming within it.

All particles are constantly moving for they are literally trapped cyclical motion. Thus the totality of All things (the first medium/ ZP energy), when moving, is a self-contained cycle, therefore it is also a "particle". ZP energy is thus the Only Particle, the Only Medium, and the Only Cycle, cycling in the infinite recursion as outlined. It is the "Cycle of cycles". Hence, ZP energy is the only thing composing all things. It is the most dense Prime Mover and simultaneously the most rarified form of energy. This is why it can travel freely in any system, and in any energy-density level, because it is already there. As such it will always be able to phase unify with some aspect of any energy-density level however tiny the phase link may be, moving all things along with it however so slightly, over aeons, being the "root" or "master" force which all things follow (flow with) being that is was first.

6. Points of Obstruction

Hopefully now the vast applications of the Ideal Field Model are becoming clear. Thus far, the basics of how and why systems merge has been examined. This section shall add detail to that mechanism.

When a density level merges, energy of that level can freely travel within it. Like water within water. Imagine then, a system five times the size of a water molecule, yet with the same energy density as water, traveling within the water. This system will be at rest in this energy density level, but will not be able to flow to all of the places that the water can simply due to its size. The areas it gets caught in, such as in a filter, are "points of obstruction". However, as this process gets smaller, such as the Nano level, it needs to be explained in terms of waveform phasing rather than particle size versus cavity size.

As mentioned, Zero Point Energy makes up all things allowing it to phase with any level of energy-density. Yet it is also the most rarified so its phasing often has subtle effects. Recalling the previous discussions on energy propagation, even ZP needs to gain energy to pass into higher density levels, but being that it is so small, the requirement is also small. It can gain this energy simply by running into itself in a loop (explained later), which is the origin of all inferior cycles (matter). So when traveling freely, ZP energy follows the easiest, most rarified path, affecting everything it contacts, ever so slightly on its return to the Prime Moving Center.

For example, fire can be easily blocked or diverted by even a flammable object. However the flow of the fire (plasma) will eventually break down the substance, carrying the barrier away bit by bit. The same with water and a mountain, the water is diverted only as long as it takes to dissolve the mountain in the path. The "teeth" of the fields of these particles of water or fire, or objects carried within them, are responsible for the disintegration of barriers, just as is the waveform of a unit of ZP energy.

When any object (bound ZP energy) is not obstructed, *free* ZP energy will normally not flow through it (unless the object is highly symmetrical), as it is far easier for the energy to flow around it. Therefore, the tiny but numerous "teeth" of the passing ZP energy slowly entice any unobstructed object to flow with the ZP toward the Prime Center. Being of inferior composition, (more complex) such an object will eventually be carried by the zero point flux to a point of obstruction. As objects pile up at a point of obstruction, impeding the Path to Return, the easiest path for the free ZP energy may then be *through* the objects as opposed to around them.

ZP energy accomplishes this by navigating through their major and minor vortices (Image 21) of which they are composed, by phasing through each energy-density level, and flowing into the natural wave cycle occurring within that object. This energy added to the object slowly begins to alter the waveform(s) of that object. In doing this, ZP carries bits of energy that it "dissolved" closer to the interior center of that object. Those freed "bits" will travel with the ZP energy through the matrixes of vortices until the size of the vortices being passed through grows smaller than the carried bit's phasing ability, this "bottleneck" then becomes that bit's point of obstruction. Similar to a filter with finer and finer screens to infinity. Thus, recursion is again apparent as a point of obstruction within a point of obstruction.

To illustrate: Imagine a floor drain, with 1-inch hole. In a storm, with water a foot deep, all matter smaller than the hole effortlessly flows through the vortex, matter larger will not flow through, but is still effected by the flux of the vortex. Furthermore, these larger particles may be so influenced by the vortex that they make it all the way to the hole and plug it up. Either the flux stops flowing or the force of the flux breaks the object into the proper form to pass the hole. If an equilibrium is achieved where the object remains *and* the flux continued to flow, another object may arrive at the same condition. This "piling up" of impure debris is another example of energy-density propagation to the center. Yet, unlike this 2-D drain illustration, this "piling up" of attracted material is happening in 3-Dimensions.

Now imagine the same drain, but instead of water flowing through it, it is acid. Soon the drain hole widens as the acid dissolves the perimeter, allowing more acid to flow in, which in turn increases the rate of corrosion. This is an example of bits of energy being carried with the free ZP energy through a vortex, the "teeth" of these bits (or the ZP itself) happen to be of the proper phasing to mesh with the internal texture of the vortex, thus carrying away or breaking free parts of the "bound" waveform condition (energy).

Every vortex eventually gets infinitely small, through which only the Zero Point energy can flow. Thus, even very tiny and nearly "pure" matter will likewise eventually get "stuck" in a vortex, unable to pass through. Here it will eventually be broken down by the persistent erosion of the ZP energy.

Each obstruction causes the energy it obstructed to flow through its smaller vortices decreasing the area of the force yet thereby increasing its pressure and consequently, its distance. This gain of distance increases the potential for more matter to be attracted by the cycle, and the faster "dissolving" of the obstruction due to the increased force of flow. Thus, this restricted flow has now a higher energy-density granting it different attributes.

In the case of a weak flux, little or no alteration would occur to an obstructed object, but in the case of a powerful flux, diamond could be stripped to carbon atoms, and those atoms stripped to their component parts, (protons and electrons), and those particles beings stripped further and further, each being of a purer and purer nature, increasing in symmetry and simplicity, until only pure energy flows through the vortex. (It is proposed that this is the mechanics of the Hutchinson effect on the transmutation of materials, and even of objects "rising" as the environment is made more energy-dense)

To illustrate: Imagine again the drain, and a sponge getting pulled into the center. For the most part it plugs the drain, but the force of the flux may be sufficient to compress the sponge and pull water *through* it, possibly increasing the force of attraction at each of the sponge's smaller openings, whereat smaller pieces of debris will likewise get stuck. If the sponge was of weak composition, the increased force of the flow would erode it as well as the matter trapped by it.

Ideally, Obstructions eventually totally collapse or collapse to the most symmetrical form, which ultimately causes higher overall symmetry and stability, increasing in perfection from the center of a system outwards. Thus in the center of a system would be the most symmetrical object. This purer and purer energy of increasing energy-density collecting in the center and progressing in symmetry, becomes geometrically more simple by its very nature. The degree of "core symmetry" is the degree to which a system behaves as the Ideal Field Model. The level of magnitude of this symmetry determines the efficiency and force of the flow of energy, and its stability. The more energy that flows the stronger is the influence of that object upon the environment. Thus, being that energy *is* mass (E=mc2), an object with a finite amount of matter, but with the potential for energy to flow through it, therefore has the potential to gain "mass" (energy) from the environment, and thus "grow". With time, this "growth" will recreate the Ideal Field Model on the next higher order of magnitude, and the forces of its cycle will likewise increase. In this way entire galaxies are created from a microscopic initiating vortex.

Sound is an example of energy transferring through a symmetrical medium. Sound speed (efficiency of transference) changes with the density and rigidity of a medium, generally the more dense and rigid the higher the efficiency. Such as in a vacuum, sound does not even travel (0m/s), while in a diamond it travels at 12000m/s, while in air it is merely 331m/s, and in water it is 1482m/s. The efficiency of transference is based on symmetry, for the denser an object is the more efficiently packed are its units, thus the more symmetrical is its composition.

The return of energy to the Prime Mover thus takes 2 paths: one towards perfect symmetry and the other towards inferior symmetry. When the energy returns in perfect symmetry, it does not bind to anything on its direct return to the center. Conversely, when energy returns imperfectly, it does not make a direct path to the center and becomes "Bound", forming particles. These particles must be then carried by free returning energy to a point of obstruction to be dismantled back into purer energy, which may then continue on to the center.

This interaction of Free and Bound ZP Energy creates "Inferior" and "Superior" energy-density. Inferior energy-density would be the familiar states of matter; temporary equilibriums of Bound energy with little degree of Free energy transferring through them. However, Superior Energy-Density is highly symmetrical geometries of bound energy, which allow for the flow through of Free ZP energy in significant degrees.

This causes *two* of each state of matter, for example solids: an "Inferior" solid would be matter arranged in a manner to keep it bound, but not necessarily well organized. The "Superior" solid (solid2) would be a super symmetrical, super "heated"(energetic), super energy-dense solid. Thus the progression of states from least to most energy-dense would be: Solid-liquid-gas-gas2 (plasma)-liquid2-solid2. These second phases are forms of "degenerate" matter, but would be better labeled as "Evolved" matter.

As the "Core symmetry" approaches these higher symmetries, not only in material geometry, but also in energy flow dynamics, other phenomena arise, such as magnetism. Magnetism is the recursive mimicking of the Ideal Field on a lesser degree of magnitude. ZP energy is flowing in the system, as the model dictates, as a tiny reflection of the Universe. It is similarly suggested that the strong force is an even higher degree of symmetry and energy-density, mimicking the Ideal Field model even more perfectly on a microcosmic scale.

It is suggested that Gravity is the result of the cycle of ZP retuning to the Prime Mover over the entire universe. Magnetism is the result of ZP returning to a cycling medium over a much smaller volume, and the strong force is the reflection of both on an even smaller scale. Each ones attractive force has a stronger "pull" per area but a proportionately shorter circumference of influence. Each being ZP energy following the Ideal Field Model in increasingly more energy dense scenarios yet on smaller orders of magnitude, they are thus recursive reflections of each other.

48

7. Boundary Condition

Thus far it has basically been discussed how waveforms of one energy-density level pass into a higher one, eventually causing the center of a system to increase in energy-density and symmetry. The mechanics of this phasing shall now be examined in even greater detail.

In a perfect Ideal field the energy-density gradient is so perfect that individual energy-density levels would be infinitely thin and therefore indistinguishable. However, in the real world, density levels are often easily distinguishable, such as where the ocean and the sky meet. This meeting place of two different energy-density levels is the place where their two waveforms meet and begin to phase unify to varying degrees, and thus it does not behave like one level or the other. It has a condition all of its own, it is the "Boundary Condition".

There are different forces that operate at the boundary of any two substances, such as the surface tension of water. Pollen particles upon water dance and spin about in Brownian motion. The pollen particles are a waveform whose energy density is between that of water and air. The pollen effectively becomes an energy-density level of its own. The action of phase unity upon it happens from the air, the water, *and* the boundary condition; thus it is pulled in all directions.

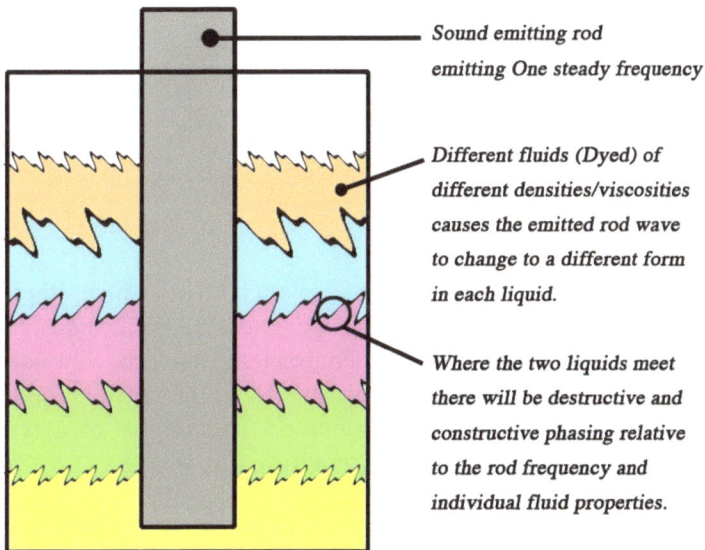

Sound emitting rod emitting One steady frequency

Different fluids (Dyed) of different densities/viscosities causes the emitted rod wave to change to a different form in each liquid.

Where the two liquids meet there will be destructive and constructive phasing relative to the rod frequency and individual fluid properties.

42.

Expanded into 3 Dimensions, the boundary condition allows waveforms to contain other waveforms without merging with them. In this form they can effectively be called "cavities" which are internal areas of less energy-density. These cavities have a specific geometry based upon the boundary condition of the two interacting waveforms, the interior and the exterior.

Such a cavity can form by a violent boundary condition between two mediums, literally stirring them up within each other, causing "wells" of little energy-density. These are similar to the "wells" in acoustic levitation where small objects are suspended in sound wells.

To illustrate: Generally a particle can merge into the field of another particle only to the level of its own acquired energy-density. Such as a steel ball which will sink to the ocean floor and a balloon will float, however, in more complex systems: If the steel ball is inside the balloon and the balloon is substantially large, the balloon will suspend the ball with it, yet if the balloon is small, then the steel ball will drag it to the bottom. However as the system falls through the various levels of energy-density, the action of the environment upon the boundary condition may be so significant that the boundary condition is broken, in other words, the ocean pressure pops the balloon and the air returns to its proper energy-density level.

Some readers may be wondering, "If energy-density propagates to the center of a system, how is it that the center of a vortex is its least dense area"? And rightly so, as it initially seems that a proposed rule is violated or a contradiction is taking place. Enough has now been discussed to illustrate this effect.

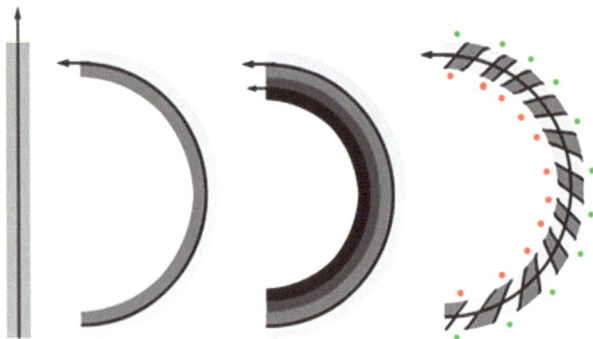

43.

When all things are moving towards one point, even if they all had ideal straight paths, the instant one point deviates *All* paths are skewed and become curves, however slight. For a unit to move in a straight line through other particles it must have incredible energy to not be deflected off of its path. This energy would be an infinite requirement, thus the Ray does not truly occur in nature, it is an Archetypal path. Thus, the curve is the natural path and all paths have some degree of curve. This degree of curve through a medium is one cause of a particle acquiring a spin due to speed differences on the two sides of the particle. (Hence, just as the Ray, Non-Spin is also an archetype).

Image 43 shows a wave path in the archetypal ray. When this path curves it immediately shortens the distance of one of its sides. Being that all electromagnetic waves propagate at the same speed, the "interior" side of the path now has the same energy in a shorter distance, thus it is more energy-dense. This curving path curves every other path that it has "cut-off" to an even sharper degree, causing them to be even more energy dense.

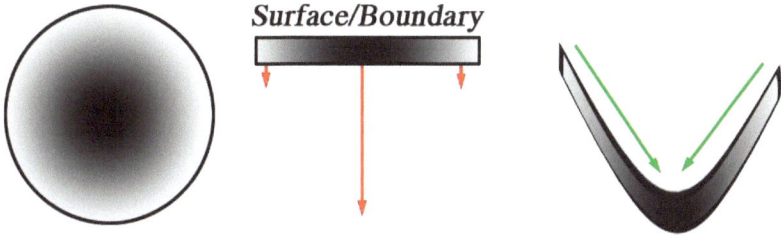

Surface/Boundary

44.

The polarized energy density of the path begins to fall through itself, attempting to equilibrate at their proper density level as they would in a sphere, thus they begin to spiral around each other. This causes the frequency of each "pair" to increase on the interior side and decrease at the perimeter, denoted by the red and green dots respectively. Thus energy density again propagates to the center, however this is still only 2-dimensional.

In a vortex on the surface of a medium, such as water, this collection of curving paths increasing in density toward the center, would basically look like the disc in image 44, with the darkest spot being the *most* energy-dense. However this disc would also be contained in a system (the ocean) and the disc's various density levels would move toward their proper density level in that system (the red arrows denote the degree of sinking to the sea floor). Thus the disc form warps, bending the boundary condition and causing the medium on the other side (the sky) to get "bottle-necked", consequently increasing in energy-density as it descends into the vortex, with increasing speed and pressure. Thus is formed the energy-density gradient of the vortex, seemingly contradicting the law of energy-density propagation at first glance.

All energy must be of equal energy-density to merge. This "bottle-neck" is one way energy-density of a medium may be raised to the significant energy level to pass the boundary condition and enter into a higher energy level. It is also a method by which a system can absorb energy from outside of the system and "grow". Thus the flow of a particle through a density gradient towards the center is synonymous with simply raising its energy. Conversely, lowering the energy-density of a field or object would cause it to *leave* the center, such as with anti-gravity.

This brings the discussion to the nature of the *ejecting* flux of the vortex. In its purest nature, it is a continuation of the recursive

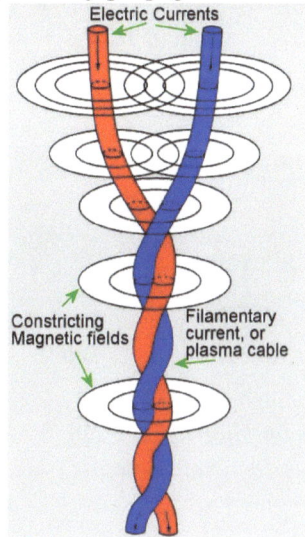

Electric Currents

Constricting
Magnetic fields

Filamentary
current, or
plasma cable

45.

51

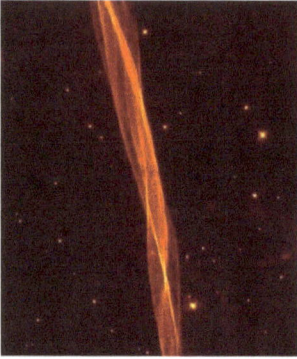

47.

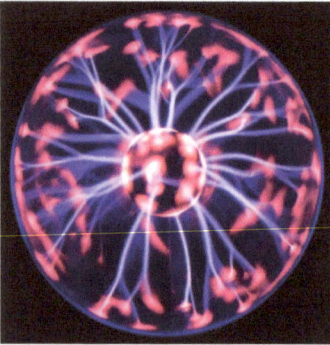

48.

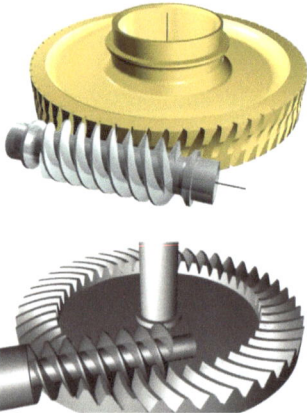

49 a.

spiral model. As discussed, density will seek its proper level, and low-density cavities formed in high-density areas will eventually be ushered to their proper level of energy density (such as light from the center of the sun). As energy accumulates in the center of a system and impure energy is broken down into purer energy, this phenomenon occurs more frequently. This low-density energy formed near the center is ejected out to its proper level. In larger systems this ejection can be quite significant as the jets of galaxy PKS 0521-36 shows.

The energy spirals out in pairs as discussed previously and shown in image 45. This pairing known as "Birkeland currents" has been recreated in the laboratory and seen in space (image 47). The fact that plasma travels in pairs can also be observed in the common "plasma ball" novelty devices (image 48).

The ejection has interesting effects on the surroundings. In addition to the force of the ejecting stream, its spiraling nature can also affect surrounding bodies. The Worm gear and Helical gear (left) represent the ejecting axis and a spiral surface of an Ideal Field system respectively. Note that a helical gear is a worm gear, just as the perimeter of the Ideal Field Model is its axis expanded. The only difference is the angle of the peaks and valleys from top to bottom. For example, the helical gears have 18 "peaks" or "teeth", but their angle is so gentle that they barely make it 20 degrees around the shaft. Now the worm gear has but one "peak", but of such a severe angle that it traverses over 2000 degrees. As can be seen, an "axis gear" could affect the top or the side of a system depending upon their flux angles. These angles of flux are illustrated in greater detail later.

52

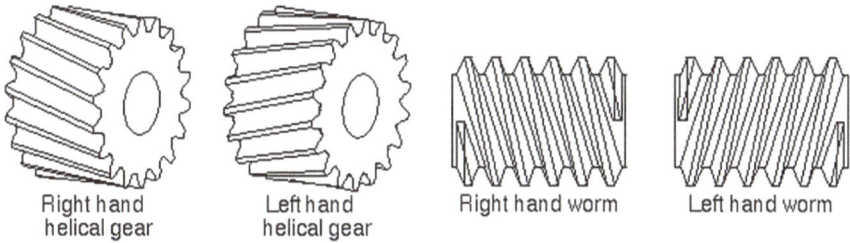

| Right hand helical gear | Left hand helical gear | Right hand worm | Left hand worm |

49 b.

Recalling the infinite recursion of the Model; *any* "ray" of flux could be said to act as an axis gear which penetrates and thus acts upon systems with the proper "receiving" geometry (a negative image of the waveform). Phenomena such as Magnetism may result when the emitted flux of a substance is received by another substance whose "receiving" geometry is the negative geometry of the spiraling flux. This very effect can be seen when fastening a board with a screw, if the board is not secured, the screw will draw the board up the screw as opposed to fastening down. This may be the mechanics of magnetic attractions. Some objects allow flux to flow through and become affected thereby, like a piece of paper on the back of a fan. Yet, other objects do not have the correct geometry, like the human hand in a magnetic field, and are completely unaffected by the passing flux.

Thus, the flux will affect only those materials whose geometry will "receive" it. This may be the very reason for aluminum's odd reactions to magnets; as it causes magnetic braking more strongly on the north side than the south. Possibly its waveforms are only favorable to one "handedness" of the flux. Similarly with bismuth's ability to repulse *both* poles of a magnet.

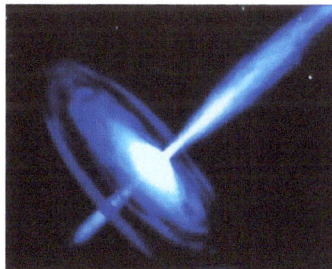

Galaxy PKS 0521-36, Hubblesite.org
The two "jets" as axis flux, which is the extended internal dual vortex of image 21,

Images: 45. Ian Tresman, plasma-universe.com 47. William P. Blair and Ravi Sankrit (Johns Hopkins University), and NASA; 48. Soenke Rahn, Wikipedia; 49. Dr. Stepan V. Lunin, spiralbevel.com

8. Models and Descriptions Thereof

Thus far, all of the basic mechanics of the Ideal Field Model have been presented with just enough information to avoid turning this paper into a technical manual. What follows are models of possible scenarios with brief descriptions, adding some additional detail, which would only have been confusing if introduced earlier. When observing the drawings, keep in mind all of the various ways to look at the model from a particle, to a field, to a system, and beyond. Furthermore, the spheres presented are merely for the ease of modeling the mechanics. To assist visualization, imagine each scenario as Ferro fluid on rotating charged spindles, and remember that they are energy-density gradients cycling through a wave function.

The Ideal Field Model grows progressively more symmetrical in the center. Ejected energy sprays out of the positive vortices and into the negative vortices while the whole model spins. These fountains are the "teeth" described earlier, and strongly resemble the arching solar flares of the sun. Note that this is simply a rough rendition; the vortices would have precise locations, and the heights of the fountains also precise, all according to the internal waveforms. Furthermore the ejecting lines shown would fan out to create a surface, much like the quintic string shown in image 35. A dynamic "gearing" system is revealed, only adding depth to the quintic string model.

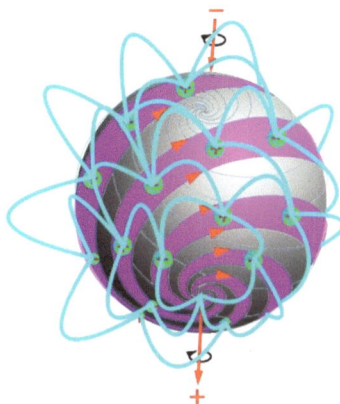

Furthermore the distance and paths of these "jets of flux" are the classical "electron orbitals". Such a model is often seen as outdated in the light of Quantum Mechanics and the uncertainty principle, suggesting not only that there are no precise orbitals, but possibly that there is no such entity as the electron. However, electromagnetic fields do form these shapes as seen in Ferro fluid, plasma labs, and galaxies. Therefore the "orbitals" are simply energy-density gradients of flux paths, and shared electrons are shared flux with certain properties defined as "electrons". Much of the "uncertainty" may arise because the "electron" does not exist, but flux does, this is why it is found everywhere.

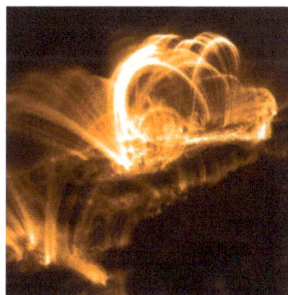

"Any discussion of the shapes of electron orbitals is necessarily imprecise, because a given electron, regardless of which orbital it occupies, can at any moment be found at any distance from the nucleus and in any direction due to the uncertainty principle. However, the electron is much more likely to be found in certain regions of the atom than in others. Given this, a boundary surface can be drawn so that the electron has a high probability to be found anywhere within the surface, and all regions outside the surface have low values. The precise placement of the surface is arbitrary, but any reasonably compact determination must follow a pattern specified by the behavior of ψ^2, the square of the wave function. This boundary surface is what is meant when the "shape" of an orbital is mentioned." – Wikipedia.

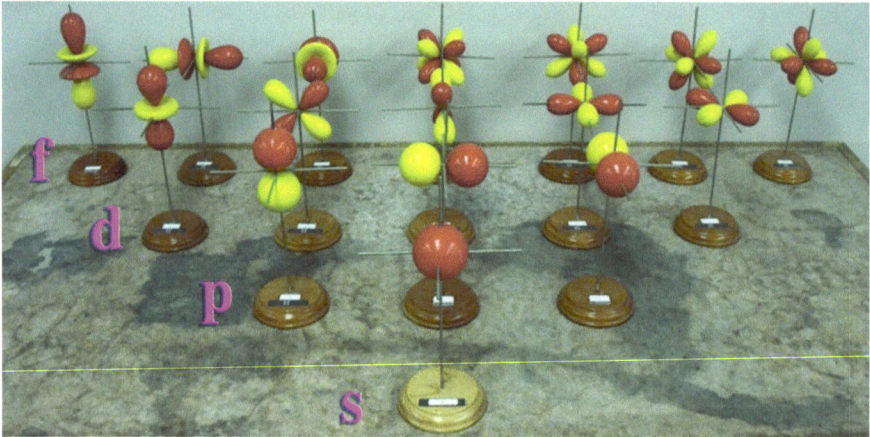

50. Louis Grace; Physics.ucsb.edu

S-Orbits: This flux configuration would be image 11 with its flow from its major axis vortices and minor surface vortices equally balanced. Much like the Sun.

P-Orbits: These would result if the flow from the major axis vortices were much greater than the minor vortices, similar to the cats eye nebulae on page 29. Even the flow of the minor vortices gets caught up in the spiraling path of the axis flux.

D-Orbits: A. The last four of this row would be a weak major axis flux with powerful minor surface vortices, appearing as a 4-arm galaxy. B. The first of this row would be that galaxy evolving to have an axis ejection.

F-Orbits: A. The first three in this row would be the result of a powerful major axis flux, with the minor vortices attracted to the flux as in the p-orbital, yet having sufficient self-attraction to avoid merging with the axis flux. Do to the significant attraction, the frequency of polarized flux eruptions from the minor vortices are shifted to the two hemispheres instead of being equally distributed across the sphere. Note that the two "disc" shapes conform roughly to the 19.47-degree anomaly of heavenly bodies shown on page 58.

B. The last four of this row adopt the dual tetrahedron geometry, which is the geometry that describes the 19.47-degree anomaly.

19.47 degrees is for a perfect sphere, which no heavenly body is. Yet note that most sunspots occur near the 19.47-degree latitudes, as do the dark bands of Jupiter and its Great Storm. Likewise, the Earth's "Bermuda triangle" is at about 23.4 degrees. Lastly, note the *hexagram* cloud formation on the Pole of Saturn as photographed by the Cassini probe.

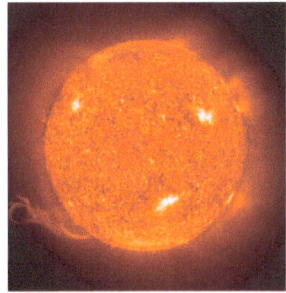

All of this geometry is caused by the tetrahedron. Why is this shape so significant? Because we know that the center of a system tends toward simplistic symmetry and the most simple form one can have in three dimensions is the tetrahedron. If a 4th dimension of Motion is added to it, it creates a dual vortex. To add motion, it must be spun. The axis of motion must be drawn through edge to edge (represented by the red dots) *not* apex to side. Apex to side creates only a single vortex, which is unbalanced with its tiny bottom and wide top.

When the tetrahedron is spun edge to edge it forms the gradient in the middle of the below image. Now when recursion is brought into play for ultimate balance (a tetrahedron spinning in a tetrahedron, spinning in a tetrahedron, etc.) the gradient on the far right is revealed. It creates a dual tetrahedron, a balanced reflection of itself as if it was spun apex-to-side one pointing up and one pointing down to achieve the balance either one could not achieve on its own. Thus this "star" is the axis of a spinning tetrahedron through multiple levels of recursion, and the tetrahedron is the simplest, most symmetrical of all 3-Dimensional shapes, exactly what the Ideal Field Model predicts for the center of evolved systems.

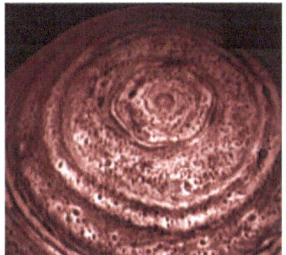

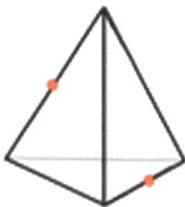

52.

Images: (3) on Right: nasa.gov

57

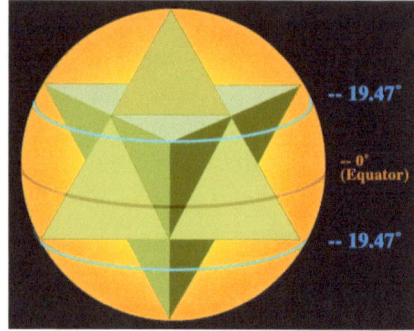

There are only two vortices, one of right hand, and one of left hand spin. In this document they are named by looking at the Positive (outward) flow. Thus, in images "a" through "m" the violet vortex is a right hand spin and the green vortex is a left hand spin. Note that the orientation of the axis is *very significant*, for there is a specific vector to it. Any vortex, when inverted, remains the same spin (as in image c), it is merely the direction of flow that *appears* to change relative to the perspective of the observer. It is important to note that "Positive" and "Negative" are simply vectors of energy flow and not "charge". "Spin" is better thought of as a cycling waveform as opposed to a spinning globe.

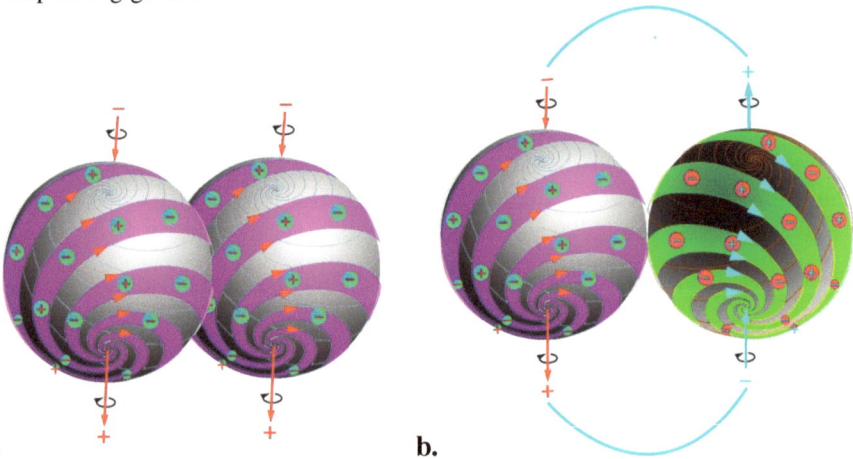

a.

b.

In classical thought, only direction of spin is taken into account, but now it can be seen that it is not just spin that identifies the particle, but the orientation of its axis. For example, the pairs in Image c. and d., both appear to be spinning towards each other, so one may initially conclude that particle "A" and "B" in image c. are equal to particles "A" and "B", in image d. Seeing these diagrams, that assumption would only be correct for the "A" particles, but very wrong for the "B" particles. This shows that direction of spin *cannot* be properly defined without knowing the orientation of the axis. Furthermore, the spin also has a *flux angle* relative to its axis due to the vortex, not a linear spin perpendicular to the axis as in classical thought.

58

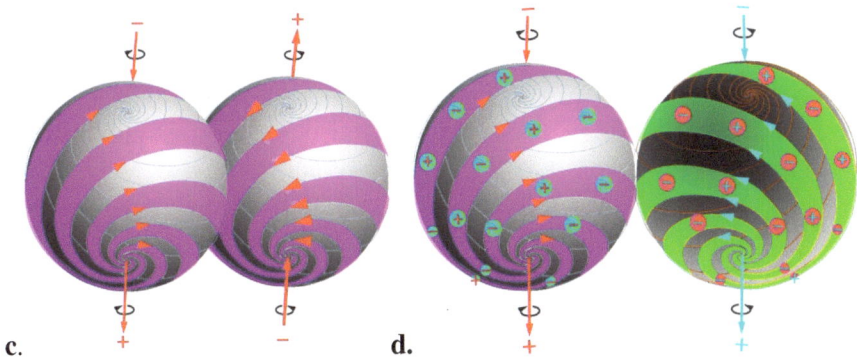

c.   +   −   d.
                           +

The small circles covering the spheres with + and − marked upon them represent the positive and negative minor surface vortices, which were represented as positive and negative "9's" in image 0. These minor vortices, when imbalanced, would form "centers of charge" cycling around the "center of mass" of the system as a whole.

There are 6 basic parameters of force that are significant in particle interaction:

1.   Relative axis orientation   2.   Spin handedness   3.   Wave pattern
4. Radius 5. Energy density and 6. Angle of interaction.   The total effects of these interacting forces directly create the properties of matter.   The following are brief examples:

Image a: Both axes oriented the same: the axes would repel, the spin would repel, the angle would repel, and only the minor vortexes at the right phase would give any hope for such a scenario to attract.

Image b: (assuming same energy density) these two particles have the same radius, and their axes are oriented to promote attraction and circulation of flux through their axes.   Thus they are pulled together by these, their strongest "charges" (flux).   Also their angle of interaction is the same, thus the "bands" overlap parallel to each other, allowing the minor vortices maximum interaction, and depending upon their phasing, will assist in either attraction or repulsion.   However, despite this, these two particles are of different handedness making the spin between them conflicting, like two grinding gears.   This presents a significant level of repulsion.   Nevertheless, the symphony of all these forces in unison may overcome to create a bond with varying degrees of *Cavities*, do to the interference of the two waveforms.

Image c: (Assuming same radii and energy density) these are two like particles, possibly two hydrogens.   These two particles are *nearly* meshed together like two gears, and their axes are aligned for flux to flow from one to the other.   However the angle of their flux cross.   The degree of this angle

59

determines the nature of the disturbance. For example, in the image below, the two "bands" cross at 90 degrees, resulting in significant repulsion, lesser angles will have less repulsion. However, the minor vortices (represented in green) may be of right phasing and strength to counteract the repulsions, possibly causing an oscillating bond. This may form a phantom vortex as in Image 12 . The right side of Image 53 shows how complex the crossing angles of flux become when 3-dimensional gradients merge. It further shows these energy-density levels being the "orbitals" of classic thought.

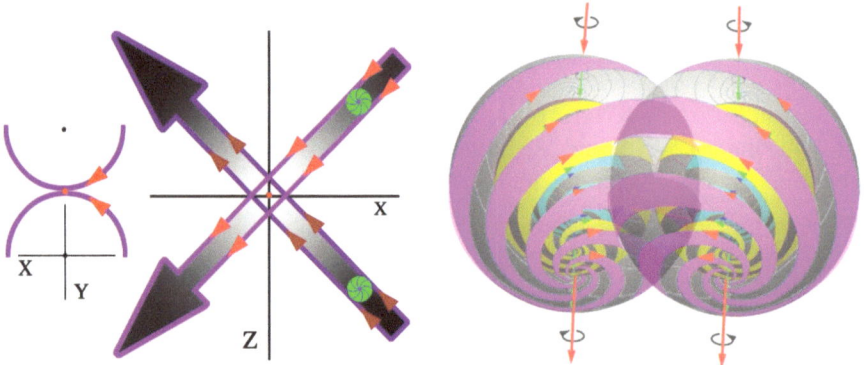

53.

Image d: (Assuming same radii and density) These two particles have opposite spin allowing them to mesh, their angle of interaction is parallel allowing for maximum meshing, and minor vortex interaction, however their axes are opposing. But as with any of these models, the attractive forces may overcome the repulsive forces resulting in a bond. As can be seen, when viewed from the Ideal Model, bonding becomes a sophisticated form of waveform interaction as opposed to simply one charge attracted to another.

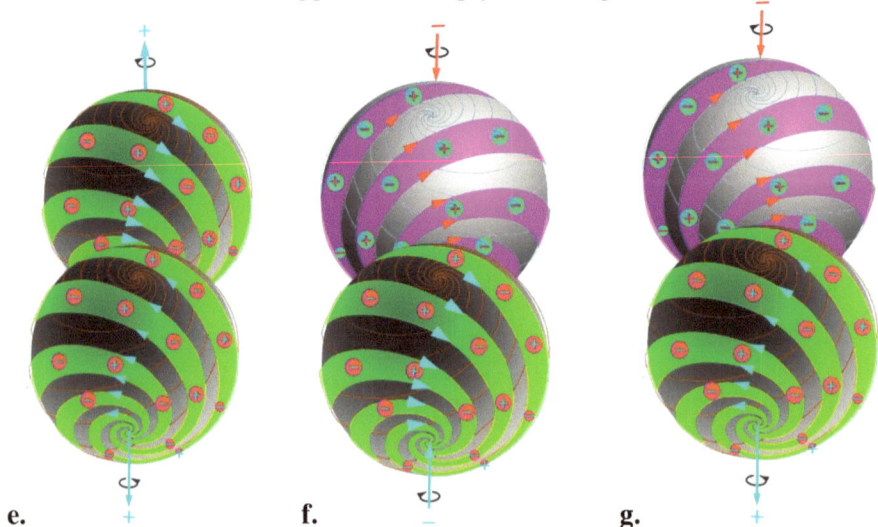

e.   f.   g.

Image e: (Assuming same radii and density) Unlike the previous models engaging side-to-side, this model engages axis to axis. It engages negative to negative, which may or may not cause an attraction. There is a chance that if it results in attraction, it will increase the vacuum of all of the negative minor vortices. However the opposing direction of the flux combined with same handedness introduces some repulsion. However the proper strength and phasing of the minor vortices may overcome that.

Image f: (Assuming same radii and density) Opposite handedness aligned with opposing positive flux would cause significant repulsion, unless the strength and phasing of the minor vortices is sufficient. The angle of interaction would be parallel allowing for the most effective interaction of the minor vortices.

Image g: (Assuming same radii and density) These two particles may have a strong axis to axis bond, but the opposing directions may cause repulsion, and this combined with opposite handedness would cause strong rapid minor vortex interactions, as they cross over each other at full parallel force. This configuration may also cause an oscillating bond if any at all.

Image h: (Assuming same radii and density) This image depicts particles of same handedness, aligned at opposite axes for strong attraction. The angle of interaction of the flux is in opposition, yet due to the same spin, it may be permanently engaged into a set of minor vortices in addition to the strong axis flow. Both sets of particles are experiencing the same forces, as mirror images.

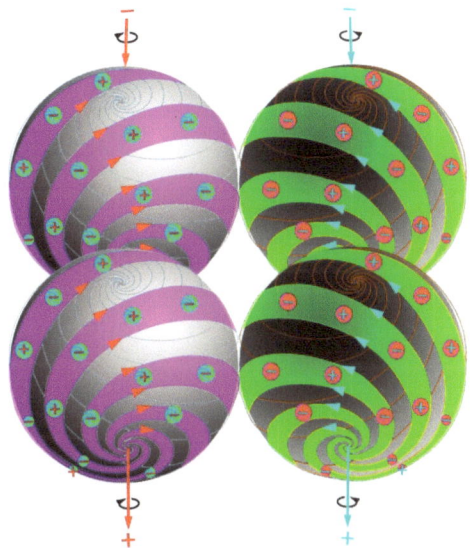

Considering the sets as units, they act as two columns of opposite handedness and mesh together well with the minor vortices aligned parallel allowing for maximum engagement. They are only repelled by their free axis both being aligned the same. Thus only the very ends would have significant repulsion, which may make this form a candidate for long protein chains.

i.

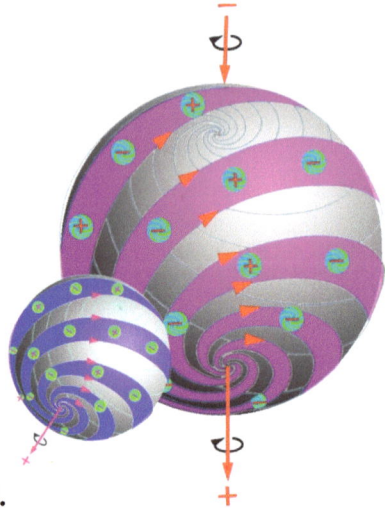

j.

Image i: (disregarding density)   This model depicts a smaller radius particle engaging a larger one axis to axis. The large amount of flux passing through the smaller particle may greatly increase the force of all of the positive minor vortices (as its energy density would be increased).   The distance that the flux of the violet particle went before bonding with the blue

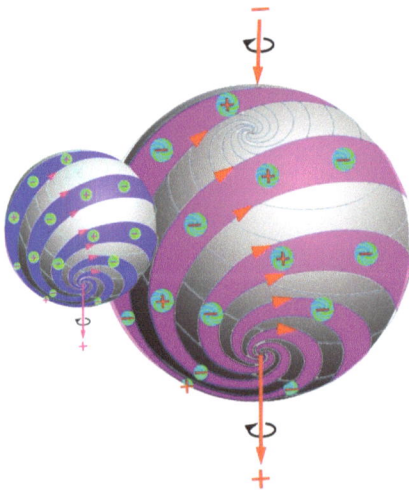

k.

particle may now be increased due to the smaller holes, furthermore it would be reaching out in different directions. This gives this particle a much greater ability to attract new matter and continue its growth.

Image j: (disregarding density) Again a smaller particle, yet this time its negative axis is bonded to a positive minor vortex of the larger particle. Its spin may be altered by the interplay of the minor vortices of both particles.   However it will be carried to the negative pole of the large particle and possibly end up as image "i".

Image k: (disregarding density) A smaller particle whose negative minor vortex is attracted to a positive minor vortex of the larger particle. This may cause the smaller particle to spin faster with increased energy density spraying a spiral out of its major axis.   It too will be short lived as it approaches the major pole of the larger particle, also possible ending as image "i" or possibly it may become image "l". The interaction of the major axes of each particle may repel keeping the small particle orbiting in the center of the larger.

62

Image l: (disregarding density) This depicts a smaller particle attracted to the positive main axis flux of a larger particle. This may cause the smaller particle to spin faster, (due to the large influx of energy), spraying out flux from its major axis as it spins. This causes greater distance and area coverage of the violet system's flux increasing its chances to attract other particles. The interplay of the minor vortices may keep this configuration stable, else it may evolve to image "i". Remember, the "spray" of this "flux" is actually just the waveform extending due to energy excitation.

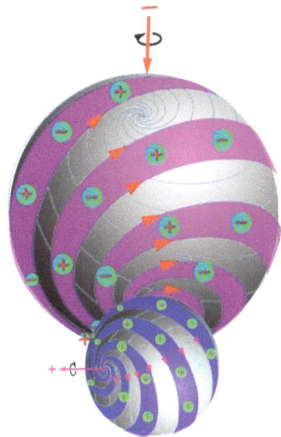

Image m: (assuming same radius and density) This image depicts two particles attracted by the major poles at 90 degrees. The flux flows from one into another, the minor vortices are in phase and of sufficient attraction to keep the objects from collapsing into an Image "h" configuration. This configuration works much like the common helical gears shown in image 54. The angle of 90 degrees shown here is merely an example, it does not intend to bar any other angles.

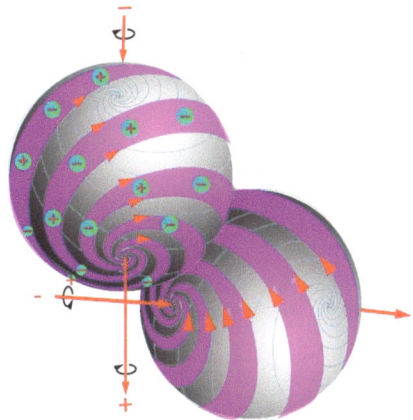

Note that exterior forces can also assist in forcing particles into any of these relationships, it need not be solely self-generated. These simple models have been shown only to illustrate the basic concepts presented in this paper. They are *far* from detailed. Note also that Major axis bonds and multiple energy-density levels may be responsible for stronger chemical bonds as opposed to just more "shared electrons". Furthermore, electronegativity is a measure of an atoms tendency towards energy-density gain, not "electron gain".

The examples presented so far have been fairly simple. In reality, the interactions would be much more sophisticated. Next are some slightly more complicated arrangements. The more complex a structure gets, the more complex the flux interactions become, and the more "characteristics" of the specific arrangement manifest.

l.

m.

Emerson Power Transmission Corp

54.

63

The following examples are of self-similar particles, such as metals. Three basic cubic structures will be touched upon. These are, from columns left to right: Simple Cubic, Body Centered Cubic (BCC) and Face Centered Cubic (FCC), each representing an increase in energy-density.

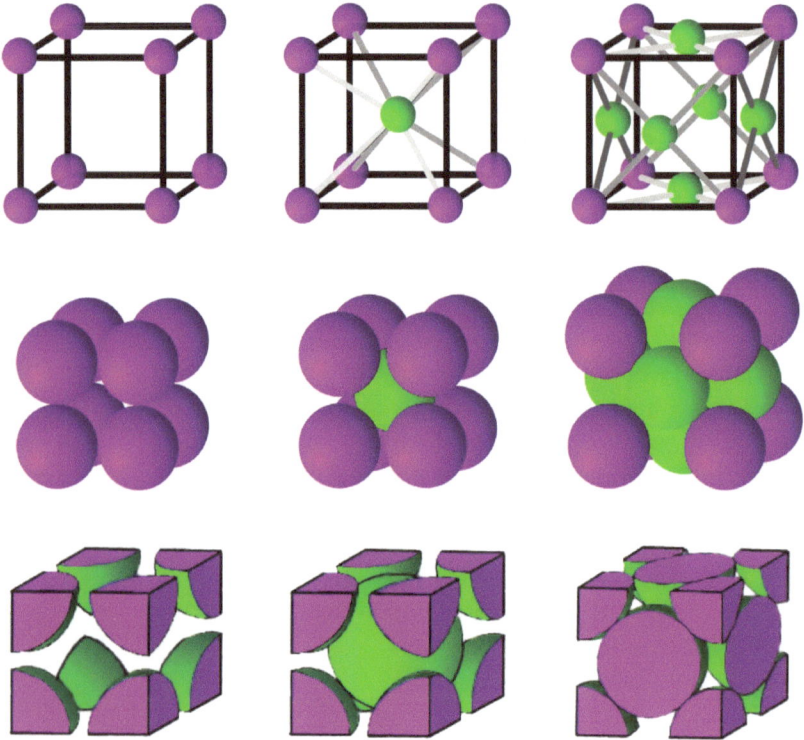

A. In a simple cubic cell each atom has 6 neighbors, is 52% pack efficient, with a void volume of 92.3% of the atom's volume. It takes 8 of these cells in a cube to create the Octahedron of NaCl.

B. The body-centered cubic has an atom between each "layer". Thus each atom has 8 neighbors, is 68% pack efficient with a void volume of 47.1% of the composing atom's volume.

C. Face-centered cubic can be thought of as a dual tetrahedron within a cube. Again this form proves to be the most symmetrical and dense with a 74% pack efficiency and a void volume of 35.1% of the composing atom's volume.

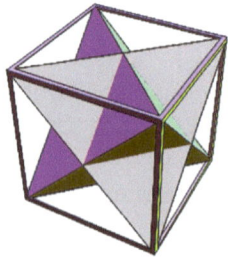

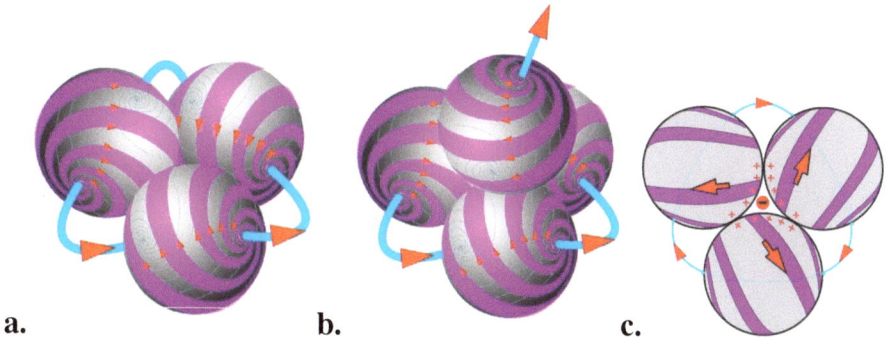

a.　　　　　　　　**b.**　　　　　　　　**c.**

Image a:　These three units are locked together axis-to-axis sharing the major axis flux in a circuit. The waveforms cycle in on one side on out on the other like a smoke ring, polarizing the respective sides, and creating an overall vortex.

Image b:　This is a system attracted to the negative side of model "a" by mechanical attraction, creating a negative center and positive edges on the other side. (Image c).

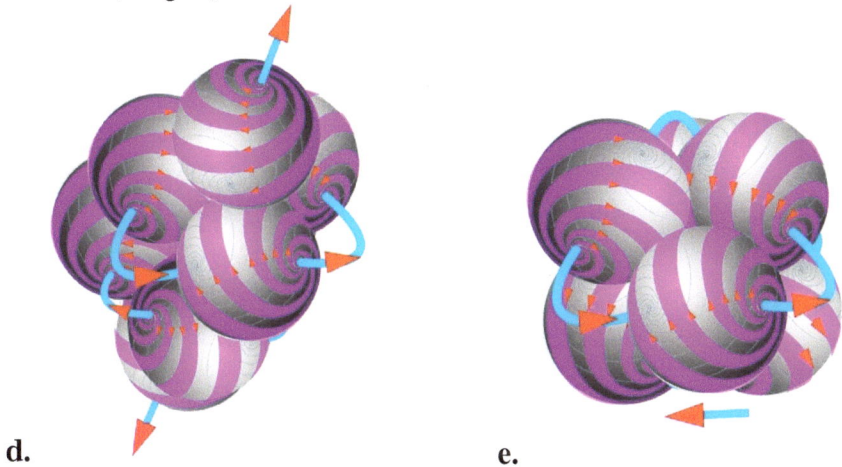

d.　　　　　　　　　　　　　**e.**

Image d:　This depicts two units of image b united by mechanical attraction.

Image e:　Adding a positive node (as image b) to this unit would be a building block of Hexagonal crystals.

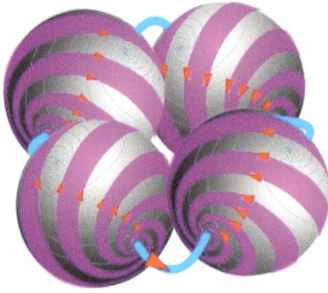

f.

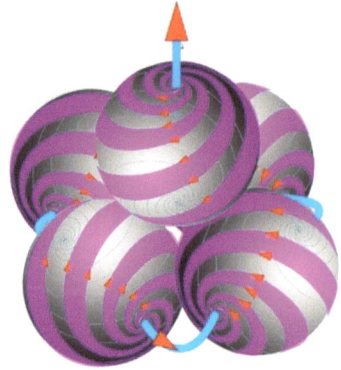

g.

Image f: This unit would be a building block of Simple cubic crystals.

Image g: This unit would be a building block of Body-Centered cubic crystals.

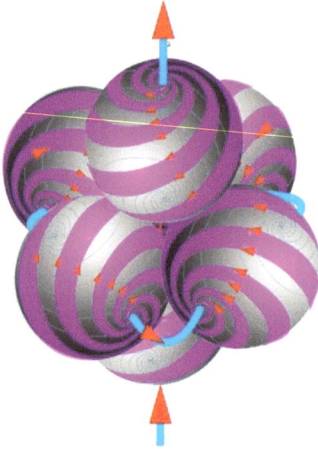

h.

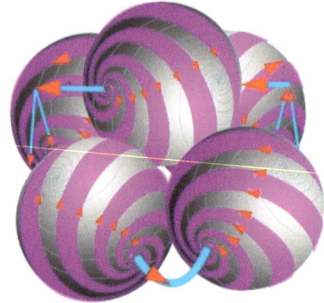

i.

Image h: This unit would be another building block of Body-Centered cubic crystals.

Image I: This unit would be another building block of Body-Centered cubic crystals.

Image j: This unit would be a building block of Face-Centered cubic crystals. (flux is ejecting "out of the page" at 60 degrees)

j.

Image k: This represents a simple cubic structure. The entire system is at equilibrium; the blue flux lines flow through the entire system uniting it into one unit, creating a phantom vortex around the substance. The energy spewing from the minor vortices travels little distance before being absorbed again – very little energy is free from the geometry of the flow lines. The bonds are strong axis to axis, but side to side the flux lines cross. This degree of repulsion, normally held in check by the surrounding forces of attraction, may be overcome by external forces.

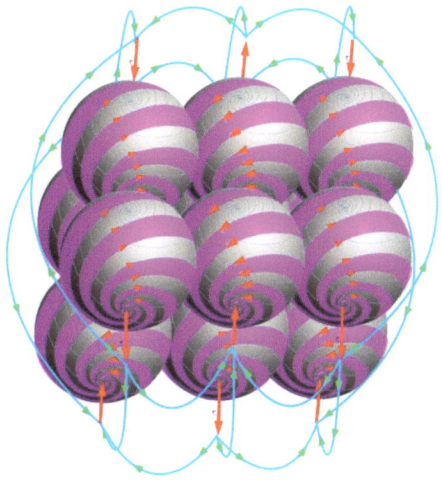

k.

Such "slip planes" could be another reason for malleability of metals, and even the liquidity of mercury. For if the angle of mercury's flux was severe (Image 53) and its strength of repulsion near equal to the axis bond, it would tend to "shear" with little or no force.

Image l: This image shows the simple cubic metal polonium, with an interesting feature, it has a cavity. The forces surrounding the cavity would cause any free energy traveling though that area to behave much like the particle of model "n" in that same location. However, unlike model "k", excess free energy can build up in this cavity without affecting the "characteristics" of the solid. If energy is added to the previous model, the energy-density of

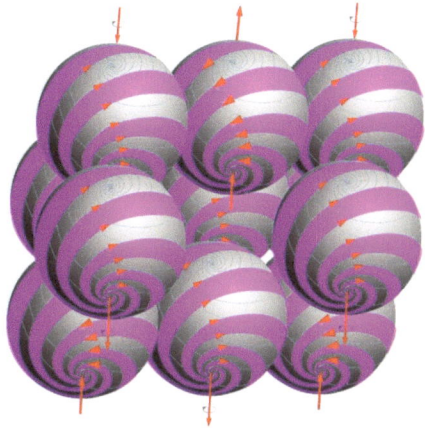

l.

each individual unit is increased and thus the characteristics of the entire substance changes – possibly manifesting as "melting" at low levels, or even polonium's high volatility. In short, there is no "detention tank" for such a flood of excess energy.

With a reservoir, as energy is pored into the system it is processed by the natural mechanics of the system, and energy beyond the processing capabilities would store up in these cavities. However, conducting energy and merely absorbing it are two different functions. Absorbing the energy is transferring the energy into the cavities and the atom itself. To conduct energy would be a function of the efficiency of the atoms ability to take the energy back out of the cavities, and the size of the individual cavities, as a cavity far too big may store energy, but may not conduct it. A cavity of a harmonic size in relation to the energy being conducted may conduct efficiently. (Such as a resonating cavity) Yet a cavity of a size that causes destructive interference, may not conduct at all regardless of its size. It is thus postulated that the shape of a cavity in a system in proportion to the composing atoms' size and energy-density, may result in higher conductive properties as the cavity acts as a resonate cavity and efficiently transmits energy. If it acts as a powerful resonating cavity, just as a magnetron, it may even emit particles, as polonium does.

Such a phantom vortex in a cavity is not ridged and quickly alters its parameters to surrounding stresses, and can "bend" to transmit energy into minor vortices of surrounding atoms. When the input of energy is withdrawn it takes time for the energy stored in the cavities to dissipate if they are non-conducting cavities, giving rise to heat storage attributes such as with ceramics.

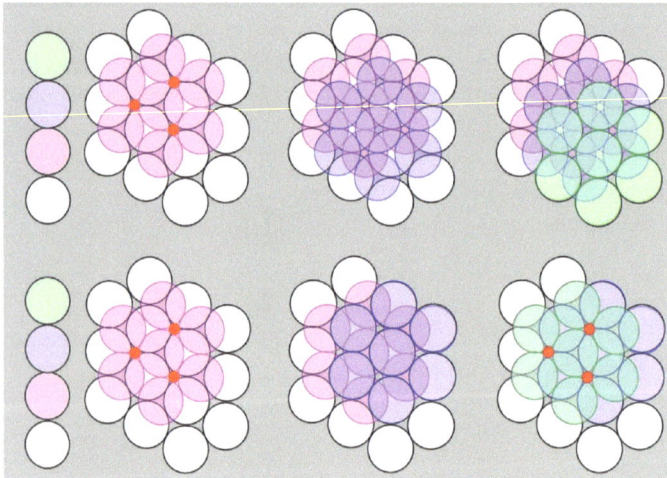

There are basically only three types of cavity, shown below. The tightest packed being the first variety, such as in Face-centered cubic and hexagonal close packed structures (left). Even though that FCC and Hexagonal are very similar, there is a critical difference when the cavities are produced.

The stacking of FCC and Hexagonal on the top and bottom respectively, follows the color code. White being the first layer and Green being the top. The red dots indicate where the "floor" (grey) can still be seen. On the final green layer of FCC, the floor can no longer be seen. However with hexagonal there remains a continuous cavity to the floor separated by tight pyramidal cavities. While the FCC, of the most

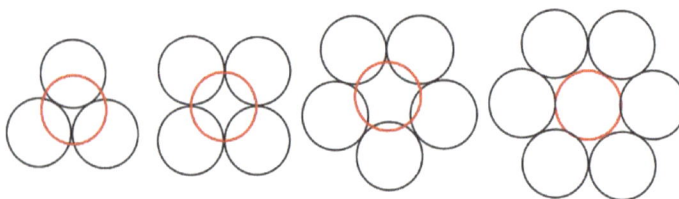

conductive metals, has only one cavity type, which is a concave dual tetrahedron apex to apex.

68

9. A Note on Time and Consciousness

A model that has infinite recursion could be discussed infinitely. Being that the model started from nothing, it is well to end here on cavities, which are likewise, "nothing". And this very statement of beginning and ending brings up time. Thus far, this entire book has been described in sequential linear time, which, in reality, does not exist. Time can only exist in relation to something else. It is a sequence of space-time events traversed at a certain rate. Therefore it is motion compared to motion, and the product of only being able to perceive a limited portion of infinity. Thus, "time" is the result of human limitations, and is not really a thing in itself.

Furthermore, there is more than one sequence of time. For a sequence to exist it has to be perceived. It is the perceiver that has the possibility to exist in a "hyper-time", for not only is there a rate of sequence, but there is also a rate of perception of that sequence. The way to prove this is with the Mobius, wherein space-time twists and repeats itself. Here the normal 4th dimensional time is skipping yet *the perceiver's* sequence of perception of the Mobius is constant, for if it were in sync with the skipping fourth dimensional time, the perceiver would be incapable of perceiving the skip. Thus the perceiver is operating on an independent sequence, which the author considers to be fifth dimensional time, or Hyper-time, the sequence of conscious perception.

Hyper-time is then the *rate of perception* of sequence, while normal "time" is the rate of the sequence itself compared to another rate of sequence. Rate is only relative to another rate, for with only one rate there is only sequence. For example: The incredibly fast movements of very small insects, such as a fly. A human rises to strike a fly, the fly appears to move lightening fast out of the way of the descending hand, and as soon as the hand slaps the table, a teakettle whistles.

The fly's total life span is only about a week long. Imagine then that the fly has the same amount of perceivable space-time events as the human does compacted into that small life span. Thus its rate of perception is countless times higher. To the fly, the human rising up from the chair may take a year of its life. The descending of the hand may take another year, the fly then mustering up the power to leap into the air and take off may take a month, etc. However, as soon as it escapes the hand that causes a titanic explosion, a teakettle whistles, which may last a decade. Thus both relative rates of sequences where the same, in proportion to the other sequences, such as the hand and the whistle, however the *rate of perception* of the sequence was quite different.

Hyper-Time can therefore be broken down into a wave applicable to the Ideal Field Model, as the pulsation of the intensity of the consciousness onto static "time-space fabric" (all possible events). The degree of **focus** being

69

amplitude, the strength of **desire** being frequency, and wavelength being **the gap** between successive space-time events. The center of the Model would be the ego sending and receiving this wave.

For example: When skiing down a hill the ego assesses the environment. This is the initial Prime Mover, the energy density of perception is strongest at the ego itself, and then thins out further away from the self, for a tree a mile away is not as important as the one right in front of the skier. The ego then compares all of the data to where it just was, and thus perceives "motion" and "forces". This is the return of the Prime Mover's force to the center (the potential void).

From here it makes a decision, which subsequently alters the next set of space-time events, such as: go left, or right. This then repeats the entire cycle. Thus the "seed" which gives impetus to the Prime Mover is the wave itself, which subsequently biases the pattern of possible space-time events to a certain set directly related to the nature of the "seed" wave, which is thought. Each time the wave returns to the center (infinite potential) it has the ability to change, which is decision. "Decision" alters the parameters of the wave to favor symmetry, simplicity and stability of accumulated "energy-density" just as dictated in the Ideal Field Model. "Energy-density" in this case is knowledge gained from experience of space-time events. Thus the consciousness "grows" according to the Ideal Field Model.

The above cycle basically constitutes "thought". Thought is thus a waveform and to communicate it to another, it must take the form of symbols, and sometimes those symbols have a corresponding sound, which is also a waveform. These special symbols are called "language". In discussing symbols of thought and how they relate to the Ideal Field Model, we come to the work of Dan Winter.

In short, Mr. Winter shows that a spiral, when lifted off of the torus and cut in half, casts shadows that generates all of the ancient Hebrew letters, and many of the letters of other sacred alphabets as well, while when uncut, it generates the Greek letters. It is interesting to note that the Ideal Field Model is a Torus, and that the Hebrew Holy book is the Torah, meaning, "The Law". Being that all things obey the Ideal Field Model, it could be said that it is A Law, If not The Law.

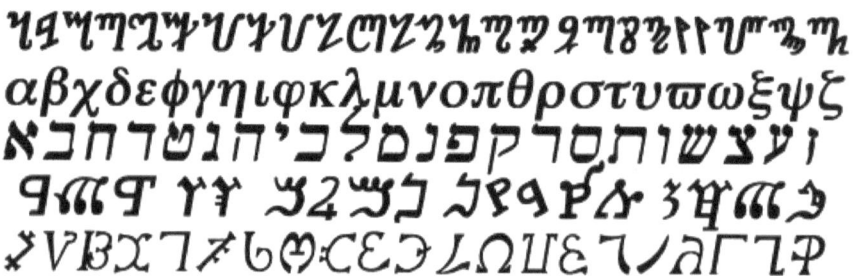

To visualize this, refer to the toroid model (Image 0.) and note the spirals thereon. Looking at the model, note only one spiral. To do this pick a colored band and imagine that you could peel it off of the torus. The shape that appears, when cut in half, would be the form in image 56b. This asymmetrical form when seen from different perspectives generates the letters.

 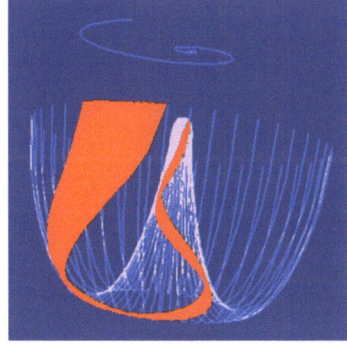

56a.   **56b.**

Dan Winter, www.goldenmean.info/dnaring/

Arizona, USA   Armenia   Guiana   New Mexico, USA   Spain

Tucson, AZ   Tyrol, Alps   United Arab Emirates   Valmonica, Italy   Venezuela

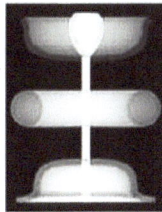 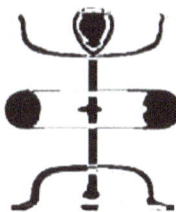

Experiment, Simulation – Derived Geometry

Dr. Anthony Peratt has collected the above ancient petroglyphs from around the globe and compared them to modern day Plasma formations created within laboratories. Note the toroidal center, axis, and dual vortex. Further research into his books and publications is strongly encouraged.

Amongst the countless significances of this, it is presented primarily to show that ancient mankind knew of this form. It is not insinuated that primitive man was producing plasma, but observing it in the ancient sky as a by-product of the cosmic electrical equilibration of our early solar system. It can reasonably be concluded that our species in the past at least knew of plasma and electromagnetism, whether or not they understood it.

Ancient man has long placed supernatural qualities on certain symbols, in light of what has been presented; their claims are not wholly unfounded. Image 56 shows that many of these 2-D letter-symbols are representations of 3-D waveforms directly from the torus. As this document has discussed, all things are 3-D waveforms following the dynamics of the Ideal Field to more or less degrees of perfection. It is the *imperfections* that lead to the characteristics of matter that we are familiar with. Precise application of waveforms to achieve desired effects may indeed seem "supernatural" to a primitive culture, as the power of language and writing must have been to early humans.

It is known that a wave's "shape" alters another wave's shape constructively or destructively. Imagine then that there was an Ideal Field (which is also a wave for it is propagating energy). Anything within the field would be affected by it. Thus generating a specific wave contrary to the field's wave geometry at a certain point would negate the effects of the field at that point. Each area in a spiraling vortex field would require a different "cancellation geometry" due to the complex vectors at each radii from the system's center.

The "AMOEBA" device of Akishima Laboratories (below) uses 50 wave generators to draw pictures and text in a pool of water by their net destructive/constructive interference. This shows that if a field of a certain vibration pattern can create a shape, then a shape can cause a field of a

certain vibration pattern. This may be the reason behind the reports of "strange writing" reported to cover the hulls of many sighted UFOs. For a charged hull would create a uniform field, but specific etchings in the hull would subsequently alter the field. For according to the Ideal Field Model, to ascend and descend in a field, an object merely has to lose or gain energy-density respectfully. However, to navigate precisely may require more complex field manipulations.

And with that otherworldly application for the Ideal Field Model this document comes to an end. The author hopes that this has sparked significant curiosity within readers to pursue the study of this model in greater detail, apply it to their works, and to ultimately expand upon it. With time, building an enlightened society of expanded consciousness with all schools of thought more easily comprehended by all being based on One Model.

~ Life blossoming as the electromagnetic field, and the spiraling ancient yin yang revealed in Ferro fluid ~

"Kaldari", wikipedia.org   Peter Terren, Tesladownunder.com

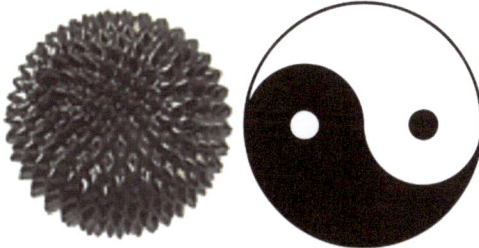

NOTE: As stated in the beginning, this book is intended to be merely a primer. A much larger and more detailed volume has yet to be released to the public. The discerning reader can grasp the various topics hinted at which will be discussed in detail in the forthcoming text.

~ IMAGE CREDITS ~

There are three groups of images used in this work:

1. Created By the Author
2. Used with Permission
3. Wikipedia Creative Commons (CC BY 3.0)

- All images used with permission or Creative Commons are credited directly below the image according to the preferences of the image creator.

- All Wikipedia.org Images are licensed under the Creative Commons Attribution 3.0 Unported license. (CC BY 3.0)

- All Other Images were created by the Author

This has been a

MASS MEDIA PUBLICATION

To Contact the Publisher or to have mail forwarded to the author, please write to:

PR@MassMedia.Space

Thank You For Your Purchase!

&

Please Visit MassMedia.Space for more products and services!

www.ingramcontent.com/pod-product-compliance
Lightning Source LLC
Chambersburg PA
CBHW041714200326
41519CB00001B/160